Synthesis Lectures on Power Electronics

This series publishes short books on topics related to power electronics, ancillary components, packaging and integration, electric machines and their drive systems, as well as related subjects such as EMI and power quality. Each Lecture develops a particular topic with the requisite introductory material and progresses to more advanced subject matter such that a comprehensive body of knowledge is encompassed. Simulation and modeling techniques and examples are included where applicable.

Imen Nouira · Bassem El Badsi

Control Strategies of Electric Drives

MATLAB/Simulink Modeling and dSPACE 1104 Validation: A Practical Approach

 Springer

Imen Nouira
University of Sfax
Sfax, Tunisia

Bassem El Badsi
University of Sfax
Sfax, Tunisia

ISSN 1931-9525   ISSN 1931-9533 (electronic)
Synthesis Lectures on Power Electronics
ISBN 978-3-031-81331-3   ISBN 978-3-031-81332-0 (eBook)
https://doi.org/10.1007/978-3-031-81332-0

This Springer imprint is published by the registered company Springer Nature Switzerland AG
The registered company address is: Gewerbestrasse 11, 6330 Cham, Switzerland

If disposing of this product, please recycle the paper.

Preface

In recent years, the rapid evolution of technology in the field of electric drives has paved the way for innovative control strategies that enhance performance, efficiency, and reliability. This book, "*Control Strategies of Electric Drives: MATLAB/Simulink Modeling and dSPACE 1104 Validation: A Practical Approach*," is designed to provide a comprehensive understanding of the fundamental concepts, methodologies, and practical applications of advanced control techniques for electric drives.

The motivation behind this work stems from the growing demand for skilled professionals who can navigate the complexities of electric drive systems. With the increasing prevalence of electric vehicles, renewable energy sources, and industrial automation, a solid foundation in control strategies has become essential. This book aims to bridge the gap between theory and practice, offering readers not only the theoretical frameworks necessary for understanding electric drives but also hands-on experiences through practical simulations and validations.

Structured into three main chapters, this book begins with an exploration of Rotor Flux Oriented Control (RFOC) for three-phase induction motors, delving into the principles of Park transformation and various implementation strategies. The subsequent chapter focuses on Direct Torque Control (DTC), providing insights into its underlying principles, advantages, and comparative analyses of simulated versus experimental results. The final chapter introduces Direct Power Control (DPC) of grid-connected DC/AC converters, highlighting the importance of rectifier topologies and power management strategies.

Through the integration of MATLAB/Simulink modeling and dSPACE 1104 validation, this book equips readers with the tools and methodologies necessary for the effective design and implementation of control strategies in electric drives. The practical examples and simulation results presented herein serve to reinforce theoretical concepts and demonstrate their applicability in real-world scenarios.

I hope that this book will serve as a valuable resource for students, researchers, and professionals in the fields of electrical engineering and control systems. It is my belief that a strong understanding of these control strategies will empower the next generation of engineers to contribute to the ongoing advancements in electric drive technology and its applications in sustainable energy solutions.

Sfax, Tunisia   Dr. Imen Nouira
October 2024   Prof. Bassem El Badsi

Contents

List of Figures

List of Tables

Rotor Flux Oriented Control of 3-Φ Induction Motor

1.1 Introduction

The scalar V/f (named also V/Hz) strategy applied to the induction machine (IM) is able to provide speed variation but does not handle transient condition control and is valid only during a steady state. This method is most suitable for applications without the need for high accuracy of speed control and leads to over-currents and over-heating.

To achieve better dynamic performance, a more complex control scheme needs to be applied, to control the IM. With the mathematical processing power offered by the micro-controllers, it is possible to implement advanced control strategies, which use mathematical transformations in order to decoupling the torque generation and the flux in an IM. Such decoupled torque and magnetization control is commonly called rotor flux oriented control (RFOC) strategy (also called vector control strategy) [1].

In order to understand the principle of the RFOC strategy, let us start with an overview of the separately excited DC motor. In this type of motor, the excitation for the stator and rotor is independently controlled. An electrical study of the DC motor shows that the produced torque and the flux can be independently tuned. Indeed, the expression of the electromagnetic torque T_{em} of a DC motor is defined as follows:

$$T_{em} = K\phi_f I_a = K'I_f I_a \tag{1.1}$$

According to Eq. (1.1), the strength of the field excitation (the magnitude of the field excitation current I_f) sets the value of the flux ϕ_f. The current I_a through the rotor windings determines how much torque T_{em} is produced. IMs have only one source that can be controlled which is the stator currents. Therefore, flux and torque depend on each other. The goal of the RFOC of IM is to be able to separately control the torque producing and magnetizing flux components to imitate the DC motor operation.

I. Nouira and B. El Badsi, *Control Strategies of Electric Drives*, Synthesis Lectures on Power Electronics, https://doi.org/10.1007/978-3-031-81332-0_1

1.2   Principle of *Park* Transform

The analytical resolution of the phase model is quite difficult because of many nonlinearities that are included there. For this reason, it is usually interesting to make variable changes leading to comprehensive models where their analytical studies are feasible.

The most popular variable change in control is the one that was proposed by *Park* [2–4]. Such a change of variable is based on the substitution of the circuits of stator and rotor by equivalent orthogonal windings, as shown in Fig. 1.1.

The *Park* transform (noted $P(\beta)$) enables the expression of the phase model (noted M_{abc}) in terms of the *Park* model (noted M_{dqo}), such as:

$$M_{abc} = P(\beta)\, M_{dqo} \tag{1.2}$$

with:

$$P(\beta) = \sqrt{\frac{2}{3}} \begin{bmatrix} \cos\beta & -\sin\beta & \frac{1}{\sqrt{2}} \\ \cos\left(\beta - \frac{2\pi}{3}\right) & -\sin\left(\beta - \frac{2\pi}{3}\right) & \frac{1}{\sqrt{2}} \\ \cos\left(\beta + \frac{2\pi}{3}\right) & -\sin\left(\beta + \frac{2\pi}{3}\right) & \frac{1}{\sqrt{2}} \end{bmatrix} \tag{1.3}$$

where: $\beta = \theta_s$ for the stator circuits, and $\beta = \theta_r$ for the rotor circuits.

The inverse matrix $P^{-1}(\beta)$ is expressed as follows:

$$M_{dqo} = P^{-1}(\beta)\, M_{abc} \tag{1.4}$$

Fig. 1.1 Relative positions of the magnetic axis with respect to those of the *dqo*-frame

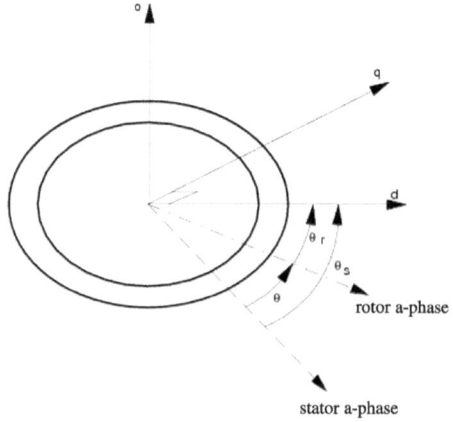

with:

$$P^{-1}(\beta) = P^{T}(\beta) = \sqrt{\frac{2}{3}} \begin{bmatrix} \cos \beta & \cos \left(\beta - \frac{2\pi}{3} \right) & \cos \left(\beta + \frac{2\pi}{3} \right) \\ -\sin \beta & -\sin \left(\beta - \frac{2\pi}{3} \right) & -\sin \left(\beta + \frac{2\pi}{3} \right) \\ \frac{1}{\sqrt{2}} & \frac{1}{\sqrt{2}} & \frac{1}{\sqrt{2}} \end{bmatrix} \quad (1.5)$$

For reasons of reducing the number of the *Park* model parameters, the *dqo*-reference frame is often linked either to the stator, to the rotor, or to the rotating field [5, 6].

1.3   *Park* Model of IM for Different Control Frame

By applying the *Park* transform to the stator and rotor equations and excluding the zero sequence components (noted "*o*"), the following equation systems are defined:

- Voltage equations:

$$\begin{cases} v_{ds} = r_s i_{ds} + \frac{d}{dt} \phi_{ds} - \omega_s \phi_{qs} & (a) \\[2mm] v_{qs} = r_s i_{qs} + \frac{d}{dt} \phi_{qs} + \omega_s \phi_{ds} & (b) \\[2mm] 0 = r_r i_{dr} + \frac{d}{dt} \phi_{dr} - \omega_r \phi_{qr} & (c) \\[2mm] 0 = r_r i_{qr} + \frac{d}{dt} \phi_{qr} + \omega_r \phi_{dr} & (d) \end{cases} \quad (1.6)$$

- Flux equations:

$$\begin{cases} \phi_{ds} = l_s i_{ds} + M i_{dr} & (a) \\[2mm] \phi_{qs} = l_s i_{qs} + M i_{qr} & (b) \\[2mm] \phi_{dr} = M i_{ds} + l_r i_{dr} & (c) \\[2mm] \phi_{qr} = M i_{qs} + l_r i_{qr} & (d) \end{cases} \quad (1.7)$$

The substitution of the phase quantities by those of *Park* in the electromagnetic torque expression gives the following relations:

$$
\begin{cases}
T_{em} = P \ (\phi_{ds}i_{qs} - \phi_{qs}i_{ds}) & (a) \\[2mm]
T_{em} = P \ (\phi_{qr}i_{dr} - \phi_{dr}i_{qr}) & (b) \\[2mm]
T_{em} = P \dfrac{M}{l_r} \ (\phi_{dr}i_{qs} - \phi_{qr}i_{ds}) & (c) \\[2mm]
T_{em} = P \dfrac{M}{l_s} \ (\phi_{qs}i_{dr} - \phi_{ds}i_{qr}) & (d)
\end{cases}
\tag{1.8}
$$

1.3.1 Control Frame Linked to the Stator Flux

Referring to the IM *Park* model where the dq-reference frame is linked to the rotating field with the d-axis is aligned on the stator flux vector ($\phi_{ds} = \phi_s$ and $\phi_{qs} = 0$), the equation systems (1.6) and (1.7) of voltages and flux become:

$$
\begin{cases}
v_{ds} = r_s i_{ds} + \dfrac{d}{dt}\phi_{ds} & (a) \\[2mm]
v_{qs} = r_s i_{qs} + \omega_s \phi_{ds} & (b) \\[2mm]
0 = r_r i_{dr} + \dfrac{d}{dt}\phi_{dr} - \omega_r \phi_{qr} & (c) \\[2mm]
0 = r_r i_{qr} + \dfrac{d}{dt}\phi_{qr} + \omega_r \phi_{dr} & (d)
\end{cases}
\tag{1.9}
$$

and:

$$
\begin{cases}
\phi_{ds} = l_s i_{ds} + M i_{dr} & (a) \\[2mm]
0 = l_s i_{qs} + M i_{qr} & (b) \\[2mm]
\phi_{dr} = M i_{ds} + l_r i_{dr} & (c) \\[2mm]
\phi_{qr} = M i_{qs} + l_r i_{qr} & (d)
\end{cases}
\tag{1.10}
$$

Also, the expression of the electromagnetic torque given by (1.8(a)) is reduced to:

$$
T_{em} = P\phi_{ds}i_{qs}
\tag{1.11}
$$

In order to imitate the separately excited DC motor operation, it is necessary to separate the control of the torque and the flux. The expression of the electromagnetic torque defined by Eq. (1.11) demonstrate that if the stator flux $\phi_s = \phi_{ds}$ is kept constant equal to its rated value, the torque T_{em} can be controlled through the q-component of the stator current i_{qs}.

Nevertheless, it is firstly fundamental to check that the stator flux ϕ_{ds} is independent of the q-component of the stator current i_{qs}. To do that, let us develop the relationship between the stator flux ϕ_{ds} and the stator currents i_{ds} and i_{qs}.

According to Eq. (1.10(a)), i_{dr} is expressed with respect to ϕ_{ds} and i_{ds} as:

$$i_{dr} = \frac{1}{M}(\phi_{ds} - l_s i_{ds}) \tag{1.12}$$

Taking into account the Eqs. (1.10(b)) and (1.10(d)), the flux ϕ_{qr} can be expressed as follows:

$$\phi_{qr} = -\frac{\sigma l_s l_r}{M} i_{qs} \tag{1.13}$$

Moreover, by considering the relations (1.10(a)) and (1.10(c)), the flux ϕ_{dr} is:

$$\phi_{dr} = \frac{l_r}{M}(\phi_{ds} - \sigma l_s i_{ds}) \tag{1.14}$$

Taking into account the Eqs. (1.12), (1.13) and (1.14), relation (1.9(c)) becomes:

$$\frac{1}{l_s}\left(1 + T_r \frac{d}{dt}\right)\phi_{ds} = \left(1 + \sigma T_r \frac{d}{dt}\right)i_{ds} - \sigma \omega_r T_r i_{qs} \tag{1.15}$$

with:

$$T_r = \frac{l_r}{r_r} \tag{1.16}$$

Referring to Eq. (1.15), one can noted that the stator flux $\phi_s = \phi_{ds}$ depends on the direct and quadrature components (i_{ds} and i_{qs}) of the stator current. Therefore, by varying i_{qs} to adjust the torque, the stator flux ϕ_{ds} is inevitably disturbed and therefore the emulation of the separately excited DC motor operation is not possible without the requirement of a decoupling system. Thus, a control reference linked to the stator flux does not allow a natural decoupling between flux and torque of IM.

1.3.2   Control Frame Linked to the Rotor Flux

By maintaining the d-axis of the reference frame constantly aligned on the rotor flux vector, the component along the q-axis of such a flux would be zero, which results in:

$$\begin{cases} \phi_{dr} = \phi_r \\ \phi_{qr} = 0 \end{cases} \tag{1.17}$$

Rewriting Eqs. (1.6) and (1.7), and taking into account that $\phi_{qr} = 0$, resulting in:

$$
\begin{cases}
v_{ds} = r_s i_{ds} + \dfrac{d}{dt}\phi_{ds} - \omega_s \phi_{qs} & (a) \\[2mm]
v_{qs} = r_s i_{qs} + \dfrac{d}{dt}\phi_{qs} + \omega_s \phi_{ds} & (b) \\[2mm]
0 = r_r i_{dr} + \dfrac{d}{dt}\phi_{dr} & (c) \\[2mm]
0 = r_r i_{qr} + \omega_r \phi_{dr} & (d)
\end{cases}
\tag{1.18}
$$

and:

$$
\begin{cases}
\phi_{ds} = l_s i_{ds} + M i_{dr} & (a) \\[2mm]
\phi_{qs} = l_s i_{qs} + M i_{qr} & (b) \\[2mm]
\phi_{dr} = M i_{ds} + l_r i_{dr} & (c) \\[2mm]
0 = M i_{qs} + l_r i_{qr} & (d)
\end{cases}
\tag{1.19}
$$

In this case, the torque expression given by (1.8(c)) is reduced to:

$$
T_{em} = \frac{PM}{l_r}\phi_{dr} i_{qs}
\tag{1.20}
$$

Referring to Eq. (1.18(c)), the current i_{dr} can be expressed as:

$$
i_{dr} = -\frac{1}{r_r}\frac{d}{dt}\phi_{dr}
\tag{1.21}
$$

The substitution of the direct component i_{dr} of the rotor current, obtained from Eq. (1.21), in the expression of the rotor flux (1.19(c)) gives:

$$
\frac{l_r}{r_r}\frac{d}{dt}\phi_r + \phi_r = M i_{ds}
\tag{1.22}
$$

Thus, the rotor flux ϕ_{dr} versus the direct stator current i_{ds} is expressed by:

$$
\phi_{dr}(p) = \phi_r(p) = \frac{M}{1 + T_r p}i_{ds}(p)
\tag{1.23}
$$

where p is the *Laplace* operator.

It is clear that, with a control frame linked to the rotor flux, the electromagnetic torque T_{em} and the rotor flux $\phi_r = \phi_{dr}$ are controlled through the two independent components (i_{qs} and i_{ds}) of the stator current. Thus, the emulation of the separately excited DC motor operation is attained.

Nevertheless, it should be noted that the flux control by the d-stator current i_{ds} is not immediate. Indeed, such a control is reached at the rotor time constant $T_r = \frac{l_r}{r_r}$. In opposition, an action on the q-stator current i_{qs} leads to an instantaneous torque response.

1.4   RFOC Implementation

In what follows, the implementation of the RFOC of IM, where the control frame is linked to the rotor flux, is developed. Such an implementation is made using a classical voltage source inverter VSI (known as six-switch three-phase inverter "SSTPI") considering two PWM techniques, namely:

◇ The current-controlled PWM technique
◇ The voltage-controlled PWM technique.

For both PWM techniques, the determination of θ_s is essential for the implementation of direct and inverse *Park* transform. To do this, one can establish the relationship between the electrical angles as follows:

$$\theta_s = \int \omega_s dt = \int (\omega_m + \omega_r) dt \tag{1.24}$$

with:

- ω_m is obtained from the mechanical speed Ω_m, such that:

$$\omega_m = P\Omega_m \tag{1.25}$$

- ω_r is determined as a function of ϕ_{dr} and i_{qr} considering Eq. (1.18(d)):

$$\omega_r = -r_r \frac{i_{qr}}{\phi_{dr}} \tag{1.26}$$

Regarding Eq. (1.19(d)), one calculate ω_r with respect to ϕ_{dr} and i_{qs}, as follows:

$$\omega_r = \frac{M}{T_r} \frac{i_{qs}}{\phi_{dr}} \tag{1.27}$$

Thus, the precision with which is determined the stator angle θ_s depends on:

∗ the accuracy of the measurement of the mechanical angle θ_m, or its estimation in the case where the position sensor is removed (sensorless control strategies). It should be noted that the speed estimation techniques are also affected by the problem of the internal parameter variation of the IM.

* the precision of the measurements of stator currents i_{ds} and i_{qs}.
* the accuracy of the estimation of rotor flux ϕ_{dr} is also restricted by the problem of parametric variations in the IM.

It is interesting to notify that several techniques of adaptation of the parameters of IM are discussed in the literature, which are aimed to improve the estimation of the stator angle, the rotor flux and the rotor pulsation. These techniques are penalized by their heavy control algorithms and eventually they increase the calculation time.

1.4.1 RFOC of IM Using Current-Controlled PWM VSI

1.4.1.1 Implementation Scheme

In the present section, the implementation scheme of the RFOC of induction machine fed by inverter using a current-controlled PWM technique is presented. This technique is based on the control of the stator currents through three hysteresis controllers, where their output signals are used to control the inverter IGBTs. Figure 1.2 presents a two-level hysteresis controller.

The hysteresis controllers are characterized by their simplicity and their fast response, but they are penalized by the fact that the frequency spectra are not well-known as in the case of the PWM techniques based on a fixed switching period.

Figure 1.3 shows the implementation scheme of the vector control strategy for an IM, where the direct axis of the control frame is aligned with the rotor flux vector, and using the current-controlled PWM inverter.

Such a control system has three control loops, such that:

* the control loop of the speed Ω_m including a PI type controller, which generates the reference electromagnetic torque T_{em}^*.
* the control loop of the electromagnetic torque T_{em} including a PI type controller, which produces the reference q-stator current i_{qs}^*.

Fig. 1.2 Two-level hysteresis controller

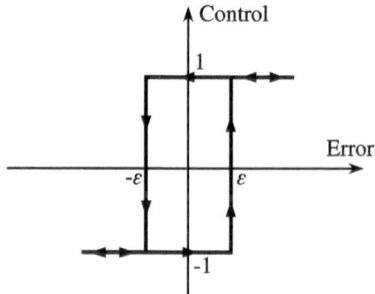

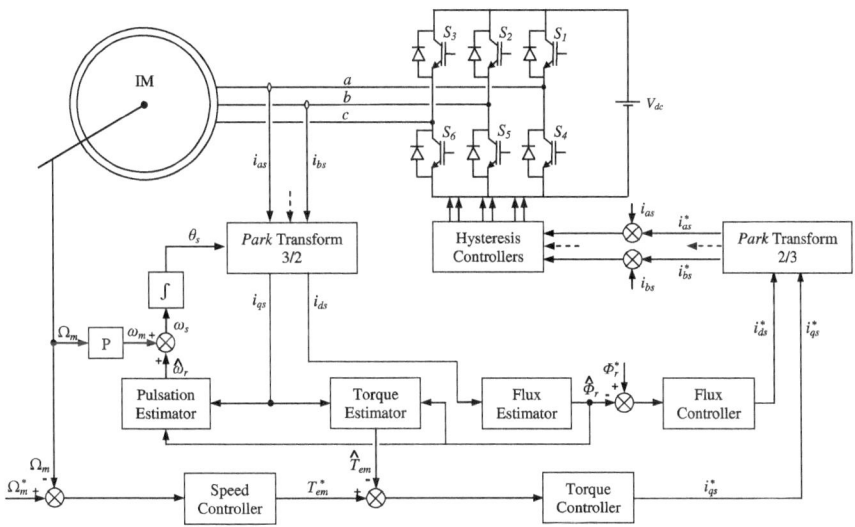

Fig. 1.3 Implementation scheme of the RFOC strategy for an IM using the current-controlled PWM VSI

* the control loop of the rotor flux ϕ_r including a PI type controller, which gives the reference d-stator current i_{ds}^*.

1.4.1.2 Synthesis of Estimators
The implementation scheme of the RFOC includes three estimators, namely:
* the flux estimator, which has one input (the d-stator current i_{ds}), gives the value of the rotor flux using the following relation:

$$\widehat{\phi}_r = \widehat{\phi}_{dr} = \frac{M}{1 + T_r p} i_{ds} \qquad (1.28)$$

* the torque estimator, which has two inputs (the estimated rotor flux $\widehat{\phi}_r$ and the q-stator current i_{qs}), produces the value of the electromagnetic torque based on the following equation:

$$\widehat{T}_{em} = P \frac{M}{l_r} \widehat{\phi}_r i_{qs} \qquad (1.29)$$

* the rotor pulsation estimator, which has the same inputs as the torque estimator (the estimated rotor flux $\widehat{\phi}_r$ and the q-stator current i_{qs}), yields the value of the rotor pulsation using the following relation:

$$\widehat{\omega}_r = \frac{M}{T_r} \frac{i_{qs}}{\widehat{\phi}_r} \qquad (1.30)$$

This expression gives an indeterminate form when the estimated rotor flux $\widehat{\phi}_r$ is zero. It is essential to add a low constant to $\widehat{\phi}_r$ in the expression of $\widehat{\omega}_r$, which is compatible with reality where the magnetic circuit comprises a residual flux.

1.4.1.3 Synthesis of Controllers

The present section treats the problem relating to the correction of the three control loops integrated in the implementation scheme of RFOC of the IM. The synthesized controllers are proportional-integral types (PI). The nonlinear system will be treated by considering the small perturbation theory which results in a linearization of the model about an operating point. Also, systems with multiple time scales will be reduced considering the singular perturbation theory.

Speed Controller

The synthesis of the speed controller is based on the type of the load. The following paragraph deals with the various load types found in industry. The mechanical equation is defined as follows:

$$T_{em} - T_l = J\frac{d\Omega_m}{dt} + f\Omega_m \qquad (1.31)$$

where: T_l is the load torque and f is the coefficient of viscous friction.

- **Constant Load Torque**: In this load characteristic, the torque is constant across the entire speed range. The power is therefore directly proportional to the speed and therefore, reducing the speed will also reduce the power required in a linear manner. For example screw compressors and conveyors are typical constant torque applications.
- **Viscous Friction Load Torque**: In this load characteristic, the torque is directly proportional to the speed. The power is proportional to square of the speed, for example rolling mills and machine tools.
- **Square-Law Load Torque**: Also called quadratic load torque, the torque of the load is proportional to the square of the speed. The load power is therefore cubically proportional to the speed, as in the case of centrifugal pumps, fans and blowers.

Constant Load Torque

Referring to Eq. (1.31), the speed loop has two inputs, such as: (i) the electromagnetic torque T_{em}, and (ii) the load torque T_l.

The synthesis of the adequate speed controller is based on the establishment of the transfer function linking the speed Ω_m and the electromagnetic torque T_{em}. The load torque T_l is considered as a disruptive input in the speed loop.

Through the *Laplace* transform, the relation (1.31) becomes:

$$T_{em}(p) = (f + Jp)\Omega_m(p) \tag{1.32}$$

The open-loop transfer function $G_{ol}(p)$ is expressed as follows:

$$G_{ol}(p) = \frac{\Omega_m(p)}{T_{em}(p)} = \frac{\frac{1}{f}}{1 + \frac{J}{f}p} \tag{1.33}$$

By inserting in the forward path, a PI controller whose transfer function could be written as follows:

$$C_S(p) = K_S \left(\frac{1 + T_S p}{T_S p} \right) \tag{1.34}$$

Thus, the open-loop transfer function $G_{ol}(p)$ becomes:

$$G_{ol}(p) = K_S \left(\frac{\frac{1}{f}}{1 + \frac{J}{f}p} \right) \left(\frac{1 + T_S p}{T_S p} \right) \tag{1.35}$$

By compensating the time constant $\frac{J}{f}$ by T_S, such that:

$$T_S = \frac{J}{f} \tag{1.36}$$

the closed-loop transfer function $G_{cl}(p)$ turns to be expressed as:

$$G_{cl}(p) = \frac{1}{1 + \frac{J}{K_S}p} \tag{1.37}$$

The proportional gain K_S is selected in the manner that the response time t_{rcl} of the closed loop system is reduced with respect to that in open loop t_{rol}:

$$\begin{cases} t_{rol} = 3\frac{J}{f} \\ \\ t_{rcl} = 3\frac{J}{K_S} \end{cases} \tag{1.38}$$

One can select the following relationship between the two response times:

$$t_{rol} = \gamma_S \, t_{rcl} \tag{1.39}$$

where γ_S is a real number greater than unity, which gives:

$$K_S = \gamma_S \, f \tag{1.40}$$

Therefore, for the case of a constant load torque the PI controller can be defined as follows:

$$\begin{cases} T_S = \dfrac{J}{f} \\[2mm] K_S = \gamma_S \, f \end{cases} \tag{1.41}$$

Load Torque Proportional to Speed

In this case, the load torque is expressed as a function of the speed as follows:

$$T_l = K_1 \Omega_m \tag{1.42}$$

The mechanical Eq. (1.31) becomes:

$$T_{em} - K_1 \Omega_m = J \frac{d\Omega_m}{dt} + f \Omega_m \tag{1.43}$$

Thus, the open-loop transfer function $G_{ol}(p)$ is:

$$G_{ol}(p) = \frac{\Omega_m(p)}{T_{em}(p)} = \frac{\dfrac{1}{K_1 + f}}{1 + \dfrac{J}{K_1 + f} p} \tag{1.44}$$

By considering a PI type controller and applying the same approach as in the case of constant load torque, the controller is defined as:

$$\begin{cases} T_S = \dfrac{J}{K_1 + f} \\[2mm] K_S = \gamma_S (K_1 + f) \end{cases} \tag{1.45}$$

where γ_S is a real number greater than unity.

Square-Law Load Torque

Considering the case of quadratic load torque, which is expressed as a function of the speed:

$$T_l = K_2 \Omega_m^2 \tag{1.46}$$

Thus, the mechanical Eq. (1.31) leads to:

$$T_{em} - K_2 \Omega_m^2 = J \frac{d\Omega_m}{dt} + f \Omega_m \tag{1.47}$$

The expression of such a load torque causes a nonlinearity. In order to synthesize the speed controller, one can apply the linearization around an operating point to the mechanical equation, and by adopting the theory of small disturbances, such that:

$$\begin{cases} T_{em} = T_{em0} + \Delta T_{em} \\ T_l \ \ = T_{l0} \ \ + \Delta T_l \\ \Omega_m = \Omega_{m0} + \Delta \Omega_m \end{cases} \qquad (1.48)$$

Therefore, the mechanical Eq. (1.31) turns to be:

$$(T_{em0} + \Delta T_{em}) - (T_{l0} + \Delta T_l) = J \frac{d}{dt}(\Omega_{m0} + \Delta \Omega_m) + f(\Omega_{m0} + \Delta \Omega_m) \quad (1.49)$$

where:

$$\begin{cases} T_{em0} = T_{l0} + f\Omega_{m0} \\ T_{l0} \ \ = K_2 \Omega_{m0}^2 \\ \Delta T_l \ \simeq 2K_2 \Omega_{m0} \Delta \Omega_m \end{cases} \qquad (1.50)$$

The open-loop transfer function $G_{ol}(p)$ is defined as follows:

$$G_{ol}(p) = \frac{\Delta \Omega_m(p)}{\Delta T_{em}(p)} = \frac{1}{(2K_2 \Omega_{m0} + f) + Jp} = \frac{\frac{1}{(2K_2 \Omega_{m0} + f)}}{1 + \frac{J}{(2K_2 \Omega_{m0} + f)}p} \qquad (1.51)$$

Thus, the parameters of the PI type controller can be selected as following:

$$\begin{cases} T_S \ = \ \frac{J}{(2K_2 \Omega_{m0} + f)} \\ K_S = \gamma_S(2K_2 \Omega_{m0} + f) \end{cases} \qquad (1.52)$$

where γ_S is a real number greater than unity.

Torque Controller

By substituting the expression of the rotor flux ϕ_{dr} given by (1.28) in the Eq. (1.20), the electromagnetic torque T_{em} is defined by:

$$T_{em} \ = \ \frac{PM^2}{l_r} \frac{1}{1 + T_r p} i_{ds} i_{qs} \qquad (1.53)$$

The application of the linearization around an operating point to Eq. (1.53), and by considering the theory of small disturbances, such that:

$$\begin{cases} T_{em} = T_{em0} + \Delta T_{em} \\ i_{qs} \ = i_{qs0} \ \ + \Delta i_{qs} \\ i_{ds} \ = i_{ds0} \end{cases} \qquad (1.54)$$

the open-loop transfer function $G_{ol}(p)$ becomes:

$$G_{ol}(p) = \frac{\Delta T_{em}(p)}{\Delta i_{qs}(p)} = \frac{PM^2 i_{ds0}}{l_r} \frac{1}{1 + T_r p} \qquad (1.55)$$

Therefore, the parameters of the PI type controller are:

$$\begin{cases} T_T = & T_r \\ \\ K_T = \gamma_T \dfrac{l_r}{PM^2 i_{ds0}} \end{cases} \qquad (1.56)$$

where γ_T is a real number greater than unity.

It is clear that the proportional gain K_T is adjusted as a function of the instantaneous value of the d-stator current i_{ds}.

Flux Controller

According to Eq. (1.28), the open-loop transfer function $G_{ol}(p)$ is expressed as follows:

$$G_{ol}(p) = \frac{\phi_{dr}(p)}{i_{ds}(p)} = \frac{M}{1 + T_r p} \qquad (1.57)$$

If the PI type controller $C_F(p)$ is defined as:

$$C_F(p) = K_F \left(\frac{1 + T_F p}{T_F p} \right) \qquad (1.58)$$

its parameters can be selected as following:

$$\begin{cases} T_F = & T_r \\ \\ K_F = \gamma_F \dfrac{1}{M} \end{cases} \qquad (1.59)$$

where γ_F is a real number greater than unity.

1.4.2   RFOC Using Current-Controlled PWM with $Matlab-Simulink$

Let us consider the simulation of the RFOC of an IM fed by a SSTPI considering the current-controlled PWM. The ratings and parameters of the IM used in the simulation are listed in Tables 1.1 and 1.2 (Fig. 1.4).

The simulation results shown in Figs. 1.5 and 1.6 deal with the start-up and the steady state of the IM, considering the following parameters:

- DC bus voltage $V_{dc} = 600$ V,
- load torque $T_l = 3$ Nm which represents 81% of the rated torque,

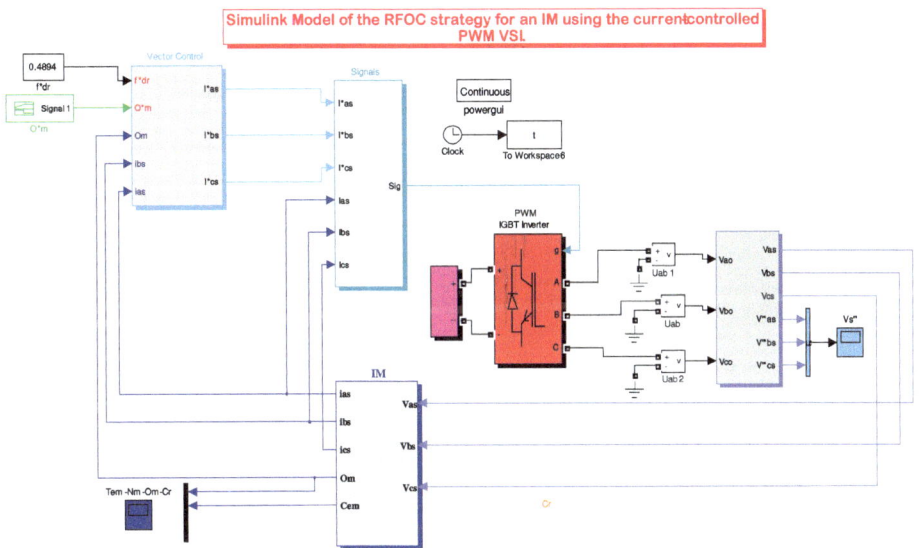

Fig. 1.4 Simulink model of the RFOC strategy for an IM using the current-controlled PWM VSI

Table 1.1 Induction machine ratings

Rated power	1.1 kW	
Efficiency	83%	
Star (voltage/current)	380 V	2.5 A
Delta (voltage/current)	220 V	4.3 A
Speed	2820 rpm	50 Hz

Table 1.2 Induction machine parameters

$r_s = 6.863\,\Omega$	$l_s = 708\,\text{mH}$	$M = 684\,\text{mH}$	$J = 0.0033\,\text{Kg.m}^2$
$r_r = 7.67\,\Omega$	$l_r = 708\,\text{mH}$	$P = 1$	$f = 0.0035\,\text{N.m.s/rad}$

- reference speed reaching $\Omega_m^* = 50\pi$ rad/s at $t = 1.5$s,
- reference rotor flux $\phi_r^* = 0.49$ Wb,
- hysteresis controllers of the stator currents with a bandwidth equal to ± 0.2 A which represents 12.5% of the rated current.

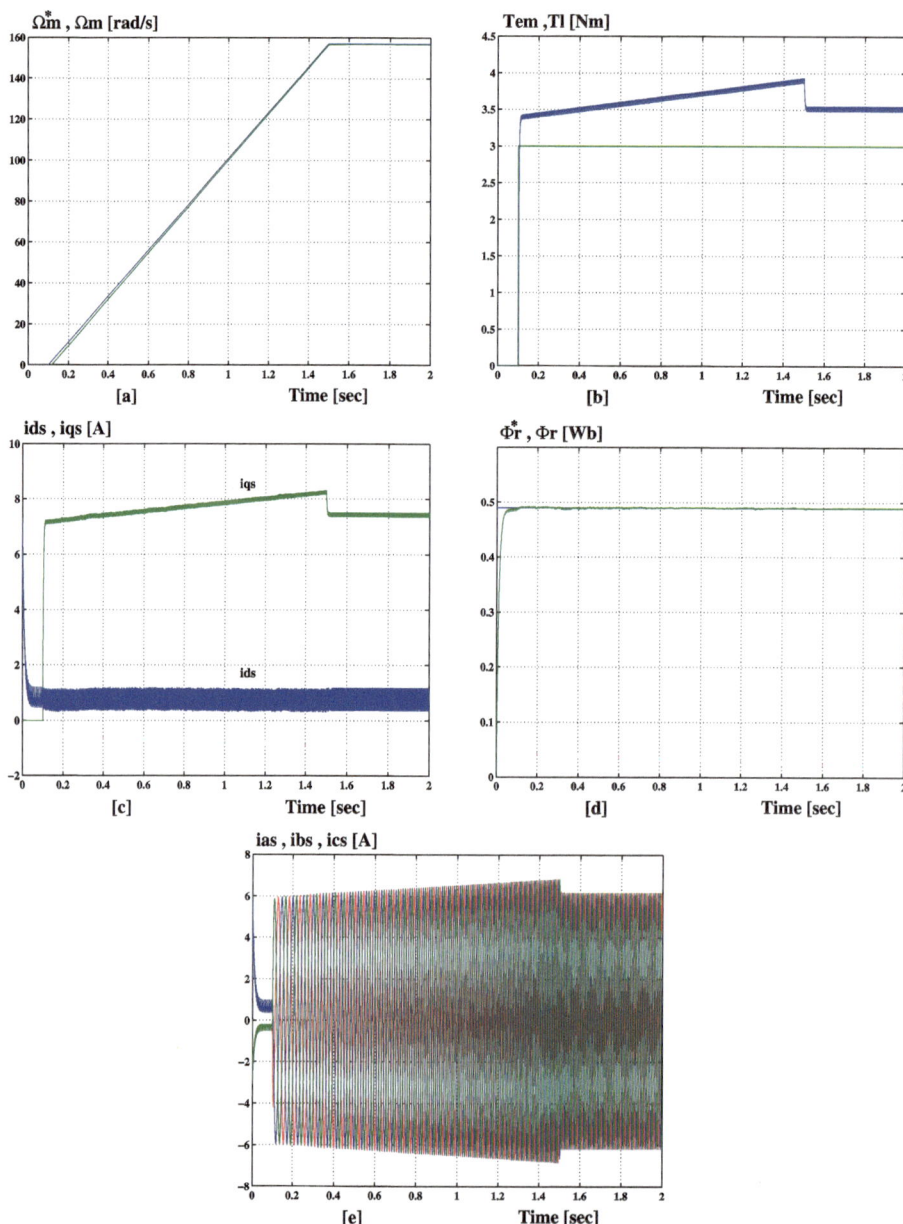

Fig. 1.5 Start-up of IM under RFOC strategy using current-controlled PWM inverter. **Legend a** Ω_m^* and Ω_m^*, **b** T_{em} and T_l, **c** i_{ds} and i_{qs}, **d** ϕ_r^* and ϕ_r, **e** i_{as}, i_{bs} and i_{cs}

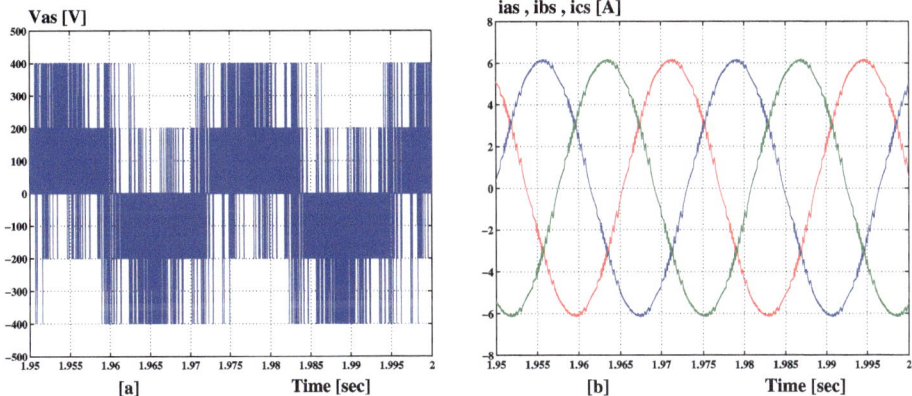

Fig. 1.6 Steady state of IM. **Legend a** V_{as}, **b** i_{as}, i_{bs} and i_{cs}

1.4.3 RFOC of IM Using Voltage-Controlled PWM VSI

1.4.3.1 Implementation Scheme

In the present section, the implementation scheme of the RFOC of IM fed by a VSI using a voltage-controlled PWM technique is presented. This technique is based on the control of the stator voltages through PWM techniques (Sinusoidal-PWM, Third Harmonic Injection-PWM, or Space Vector-PWM), where the output signals of a PWM modulator control the inverter IGBTs.

Although the implementation of RFOC requires the references of the direct and quadrature components (i_{ds}^* and i_{qs}^*) of the stator current vector, it can be developed through a voltage-controlled PWM inverter by generating the references of the direct and quadrature components (v_{ds}^* and v_{qs}^*) of the stator voltage vector.

Referring to Eqs. (1.18(a)) and (1.18(b)), and by replacing the expressions of the d- and q-stator flux components given by (1.19(a)) and (1.19(b)), the d- and q-stator voltage components are defined as follows:

$$\begin{cases} v_{ds} = r_s i_{ds} + \dfrac{d}{dt}(l_s i_{ds} + M i_{dr}) - \omega_s(l_s i_{qs} + M i_{qr}) \\[2mm] v_{qs} = r_s i_{qs} + \dfrac{d}{dt}(l_s i_{qs} + M i_{qr}) + \omega_s(l_s i_{ds} + M i_{dr}) \end{cases} \tag{1.60}$$

According to the voltage equations given by (1.60), it is clear that the d-stator voltage v_{ds} affects both on i_{ds} and i_{qs}, therefore on the rotor flux ϕ_{dr} and on the electromagnetic torque T_{em}. Also, one can noted the same remark for the q-stator voltage v_{qs} [7].

Thus, the decoupling condition between the flux and torque is no longer achieved. To overcome this problem, it is necessary to introduce two variables v_{ds}' and v_{qs}', where v_{ds}' affects only the d-stator current component i_{ds}, and v_{qs}' affects only the q-stator current component i_{qs}.

Let us start by developing the expressions of the voltages v_{ds} and v_{qs} by substituting, in the equation set (1.60), the following expressions of the d- and q-rotor current components:

$$\begin{cases} i_{dr} = \frac{1}{l_r}(\phi_{dr} - Mi_{ds}) \\ \\ i_{qr} = -\frac{M}{l_r}i_{qs} \end{cases} \tag{1.61}$$

Thus:

$$\begin{cases} v_{ds} = r_s i_{ds} + \frac{d}{dt}\left(l_s i_{ds} + \frac{M}{l_r}(\phi_{dr} - Mi_{ds})\right) - \omega_s\left(l_s - \frac{M^2}{l_r}\right)i_{qs} \\ \\ v_{qs} = r_s i_{qs} + \omega_s\left(l_s i_{ds} + \frac{M}{l_r}(\phi_{dr} - Mi_{ds})\right) + \frac{d}{dt}\left(l_s - \frac{M^2}{l_r}\right)i_{qs} \end{cases} \tag{1.62}$$

Taking into account the expression of the magnetic dispersion coefficient $\sigma = 1 - \frac{M^2}{l_s l_r}$, the equation system (1.62) becomes:

$$\begin{cases} v_{ds} = \underbrace{\left(r_s + \sigma l_s\frac{d}{dt}\right)i_{ds} + \frac{M}{l_r}\frac{d}{dt}\phi_{dr}}_{v'_{ds}} \quad \underbrace{-\ \omega_s\sigma l_s i_{qs}}_{v_{dscom}} \qquad (a) \\ \\ v_{qs} = \underbrace{\left(r_s + \sigma l_s\frac{d}{dt}\right)i_{qs}}_{v'_{qs}} \quad + \underbrace{\omega_s\frac{M}{l_r}\phi_{dr} + \omega_s\sigma l_s i_{ds}}_{v_{qscom}} \qquad (b) \end{cases} \tag{1.63}$$

Therefore, referring to Eq. (1.63(a)) the control of the rotor flux ϕ_{dr} is performed through the v'_{ds} component, instead of i_{ds} in the case of the RFOC of IM fed by the current-controlled PWM inverter. Also, accounting to Eq. (1.63(b)) the electromagnetic torque T_em is adjusted by the v'_{qs} component, instead of i_{qs} in the case of the RFOC of IM fed by the current-controlled PWM inverter.

Figure 1.7 shows the implementation scheme of the vector control strategy for an IM, where the direct axis of the control frame is aligned with the rotor flux vector, and using the voltage-controlled PWM inverter.

Like in the case of the control system shown in Fig. 1.3, the implementation scheme of Fig. 1.7 involves:

- three control loops (speed Ω_m, electromagnetic torque T_{em}, and rotor flux ϕ_r), which are based on PI type controllers,
- three estimators (rotor flux $\widehat{\phi}_r$, electromagnetic torque $\widehat{T}_{em}$, and rotor pulsation $\widehat{\omega}_r$), which are based on relations (1.28), (1.29), and (1.30).

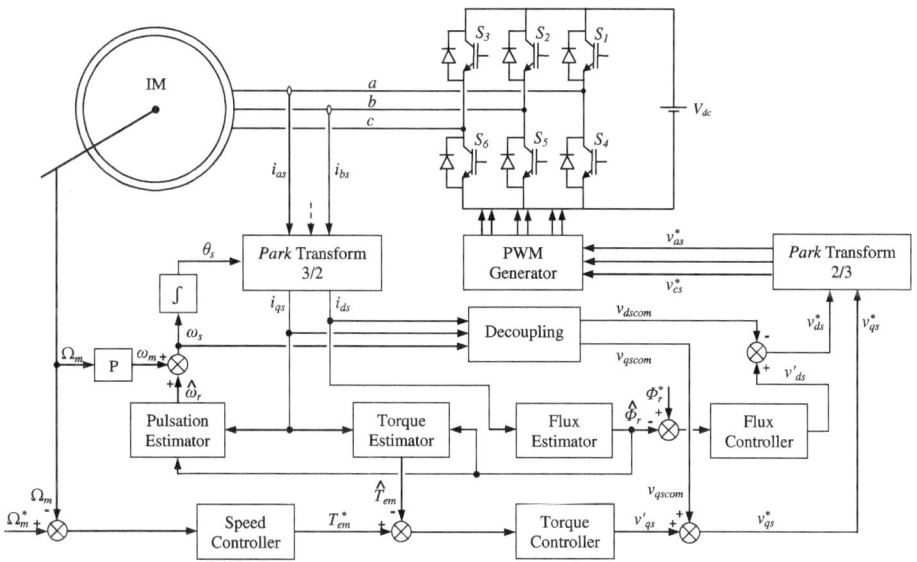

Fig. 1.7 Implementation scheme of the RFOC strategy for an IM using the voltage-controlled PWM VSI

1.4.3.2 Synthesis of Controllers
Speed Controller

The synthesis of the speed controller is carried out in the same manner as that developed in paragraph 1.4.1.3, dealing with the RFOC of IM fed by current-controlled PWM VSI.

Torque Controller

The expression of the electromagnetic torque is:

$$T_{em} = \frac{PM}{l_r}\phi_{dr}i_{qs} \tag{1.64}$$

If the rotor flux ϕ_{dr} is kept constant equal to its rated value, and by substituting the q-stator current component i_{qs} by its expression given by the relation (1.63(b)), the torque T_{em} can be controlled through the component v'_{qs}, such as:

$$T_{em}(p) = \frac{PM\phi_{dr}}{l_r}\left(\frac{v'_{qs}(p)}{r_s + \sigma l_s p}\right) \tag{1.65}$$

Therefore, the open-loop transfer function $G_{ol}(p)$ is expressed as follows:

$$G_{ol}(p) = \frac{T_{em}(p)}{v'_{qs}(p)} = \frac{PM\phi_{dr}}{l_r r_s}\left(\frac{1}{1 + \sigma T_s p}\right) \tag{1.66}$$

Thus, the parameters of the PI type controller can be selected as following:

$$\begin{cases} T_T = \sigma T_s \\ K_T = \gamma_T \dfrac{l_r r_s}{P M \phi_{dr}} \end{cases} \tag{1.67}$$

where γ_T is a real number greater than unity.

Flux Controller

Referring to (1.63(a)), the equation related v'_{ds} component, the direct stator current i_{ds} and the rotor flux ϕ_{dr} is:

$$v'_{ds}(p) = (r_s + \sigma l_s p) i_{ds} + \frac{M}{l_r} p \phi_{dr} \tag{1.68}$$

Knowing that:

$$i_{ds} = \frac{1 + T_r p}{M} \phi_{dr} \tag{1.69}$$

the relationship between v'_{ds} and ϕ_{dr} becomes:

$$v'_{ds} = \frac{r_s}{M} \left(1 + (T_s + T_r) p + \sigma T_s T_r p^2 \right) \phi_{dr} \tag{1.70}$$

where:

$$T_s = \frac{l_s}{r_s} \tag{1.71}$$

The open-loop transfer function $G_{ol}(p)$ is defined as:

$$G_{ol}(p) = \frac{\phi_{dr}(p)}{v'_{ds}(p)} = \frac{\dfrac{M}{r_s}}{\sigma T_s T_r p^2 + (T_s + T_r)p + 1} \tag{1.72}$$

which can be rewriting as follows:

$$G_{ol}(p) = \frac{\phi_{dr}(p)}{v'_{ds}(p)} = \frac{\dfrac{M}{r_s \sigma T_s T_r}}{(p - S_1)(p - S_2)} \tag{1.73}$$

with:

$$\begin{cases} S_1 = \dfrac{-(T_s + T_r) - \sqrt{T_r^2 + T_s^2 + (2 - 4\sigma)T_s T_r}}{2\sigma T_s T_r} \\ S_2 = \dfrac{-(T_s + T_r) + \sqrt{T_r^2 + T_s^2 + (2 - 4\sigma)T_s T_r}}{2\sigma T_s T_r} \end{cases} \tag{1.74}$$

Let us call:

$$
\begin{cases}
T_1 = -\dfrac{1}{S_1} \\[2mm]
T_2 = -\dfrac{1}{S_2}
\end{cases}
\tag{1.75}
$$

Accounting for that:

$$
T_1 T_2 = \frac{1}{S_1 S_2} = \frac{1}{2}
\tag{1.76}
$$

the open-loop transfer function $G_{ol}(p)$ given by Eq. (1.73) turns to be:

$$
G_{bo}(p) = \frac{\phi_{dr}(p)}{v'_{ds}(p)} = \frac{\dfrac{M}{2r_s \sigma T_s T_r}}{(1 + T_1 p)(1 + T_2 p)}
\tag{1.77}
$$

This transfer function includes two time scales, which could be reduced by neglecting the fastest dynamic (which means the smallest time constant), according to the singular perturbation theory.

Taking into account that:

$$
S_1 < S_2 < 0
\tag{1.78}
$$

which signify that:

$$
0 < T_1 < T_2
\tag{1.79}
$$

Thus, the open-loop transfer function $G_{ol}(p)$ is reduced to

$$
G_{bo}(p) = \frac{\phi_{dr}(p)}{v'_{ds}(p)} \simeq \frac{\dfrac{M}{2r_s \sigma T_s T_r}}{1 + T_2 p}
\tag{1.80}
$$

Therefore, the parameters of the PI type controller can be selected as following:

$$
\begin{cases}
T_F = T_2 \\[2mm]
K_F = \gamma_F \dfrac{2r_s \sigma T_s T_r}{M}
\end{cases}
\tag{1.81}
$$

where γ_F is a real number greater than unity.

1.4.4   RFOC Using Voltage-Controlled PWM with $Matlab-Simulink$

Let us consider the simulation of the RFOC of an IM fed by a SSTPI considering the voltage-controlled PWM technique. The ratings and parameters of the IM used in the simulation are the same presented in Tables 1.1 and 1.2.

The simulation results shown in Figs. 1.8 and 1.9 deal with the start-up and the steady state of the IM considering the following parameters:

- DC bus voltage $V_{dc} = 600$V,
- load torque $T_l = 3$Nm, which represents 81% of the rated torque, applied at $t = 1.5$s,
- reference speed reaching $\Omega_m^* = 50\pi$rad/s at $t = 1$s,
- reference rotor flux $\phi_r^* = 0.49$Wb,
- sampling period $T_s = 10^{-4}$s, which corresponds to a switching frequency $f_s = 20$kHz.

1.4.5   Analysis of Simulation Results

To interpret the simulation results of the Rotor Flux Oriented Control (RFOC) applied to an induction motor (IM) using Pulse Width Modulation (PWM) techniques, here is a detailed analysis based on the behavior observed in the simulated results.

1. **Start-Up Performance** (Figs. 1.4 and 1.5).
   - *Speed Response*: The plot shows the motor's reference speed Ω_m^* and the actual speed Ω_m over time. At start-up, the motor initially accelerates, and the speed follows the reference value, stabilizing at the steady-state speed after a brief transient period. The close tracking of the reference speed indicates that the RFOC strategy is effective in regulating the speed of the motor.
   - *Torque Response*: The electromagnetic torque T_{em} and load torque T_l are plotted. At start-up, there is a surge in electromagnetic torque as the motor works to overcome inertia and accelerate. Once the motor reaches steady-state, the torque stabilizes and matches the load torque. This behavior demonstrates the ability of RFOC to manage torque effectively during dynamic changes, such as acceleration and load application.
   - *Stator Currents* (i_{ds} and i_{qs}): The plots for ids and iqs represent the direct and quadrature components of the stator current. These currents stabilize over time, with iqs controlling the torque and ids managing the rotor flux. The smooth convergence of these values shows effective decoupling between flux and torque, which is a key feature of RFOC.
   - *Rotor Flux* ϕ_r^*: The rotor flux plot shows how the flux stabilizes at a reference value after an initial transient. Maintaining a constant rotor flux is crucial for efficient

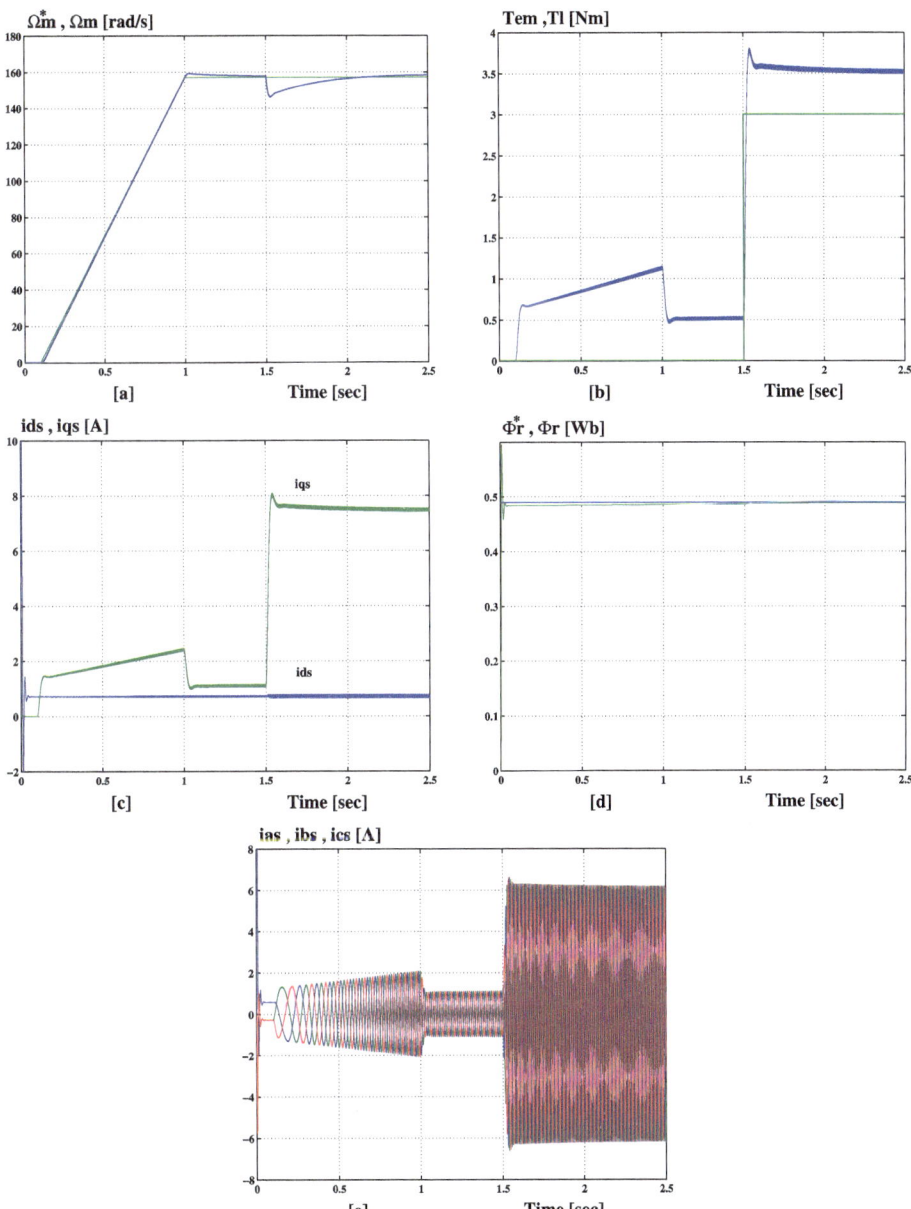

Fig. 1.8 Start-up of IM under RFOC strategy using voltage-controlled PWM VSI. **Legend a** Ω_m^* and Ω_m^*, **b** T_{em} and T_l, **c** i_{ds} and i_{qs}, **d** ϕ_r^* and ϕ_r, **e** i_{as}, i_{bs} and i_{cs}.

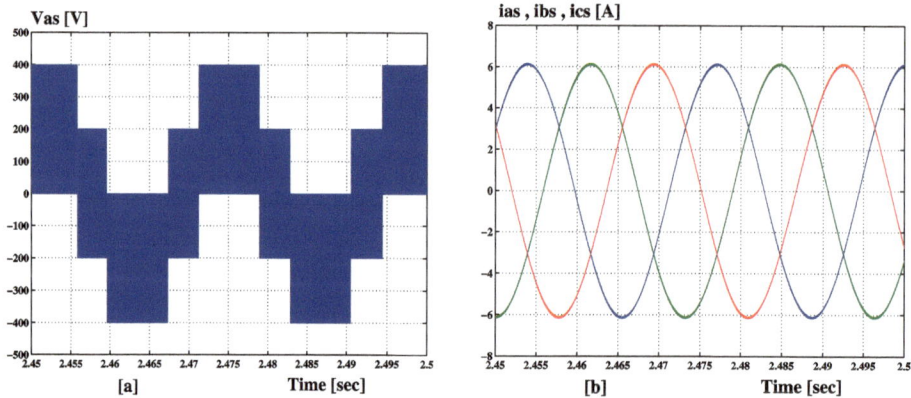

Fig. 1.9 Steady state of IM. **Legend a** V_{as}, **b** i_{as}, i_{bs} and i_{cs}

torque generation, and the simulation confirms that the flux remains steady during operation.

2. **Steady-State Performance** (Figs. 1.7 and 1.8).

   - *Voltage Response*: In steady-state, the phase voltage (V_{as}) is sinusoidal, indicating a stable operation of the PWM inverter. The absence of significant voltage fluctuations highlights the effectiveness of the voltage-controlled PWM strategy.
   - *Current Response*: The stator currents (i_{as}, i_{bs} and i_{cs}) remain balanced and sinusoidal, which reflects smooth steady-state operation with minimal current distortion. This ensures that the motor operates efficiently with reduced harmonic content in the current waveforms, which is essential for minimizing losses and heating.

Interpretation Summary: The simulation results show that RFOC, when implemented with both current-controlled and voltage-controlled PWM techniques, provides robust control of the induction motor. The start-up results confirm the strategy's effectiveness in controlling the torque and speed during dynamic conditions, with quick stabilization. The steady-state results indicate efficient motor operation with minimal flux and current ripples, ensuring optimal performance. The overall performance demonstrates that RFOC is well-suited for applications requiring precise control and quick response in both transient and steady-state conditions.

1.5   Conclusion

The Rotor Flux Oriented Control (RFOC) has been demonstrated to effectively decouple the control of torque and flux in induction motors, offering performance advantages in terms of fast dynamic response and precise control. By aligning the control frame with the rotor flux vector, RFOC ensures independent control of the torque and magnetizing components, mimicking the performance of a DC motor. Both current-controlled and voltage-controlled PWM techniques were shown to be suitable for implementing RFOC, each with its own set of advantages. Simulation results confirm the robustness of RFOC, particularly in maintaining accurate flux control and torque response under varying operational conditions. This control strategy is therefore well-suited for applications requiring high dynamic performance and precision in speed and torque control.

References

1. Yousfi, L., Aoun, S., & Sedraoui, M. (2022). Speed sensorless vector control of doubly fed induction machine using fuzzy logic control equipped with Luenberger observer. *International Journal of Dynamics and Control, 10*(6), 1876–1888.
2. Tran, C. D., Brandstetter, P., Nguyen, M. C. H., Ho, S. D., Bach, H. D., & Pham, P. N. (2020). A robust diagnosis method for speed sensor fault based on stator currents in the RFOC induction motor drive. *International Journal of Electrical and Computer Engineering, 10*(3), 3035.
3. Ouarda, A., El Badsi, B., & Masmoudi, A. (2018). Direct RFOC strategies aimed to symmetrical two-phase IM drives: Comparison between B4-and B6-inverters in the stator. *IEEE Transactions on Power Electronics, 33*(11), 9772–9782.
4. Tabasian, R., Ghanbari, M., Esmaeli, A , & Jannati, M. (2022). A novel speed control strategy for 3-phase induction motor drives with star-connected under single-phase open-circuit fault using modified RFOC strategy. *ISA Transactions, 125*, 492–513.
5. El Badsi, I. N., El Badsi, B., & Masmoudi, A. (2016). Experimental evaluation of RFOC and DTC strategies for B6-VSI fed induction motor drives. In *2016 17th international conference on sciences and techniques of automatic control and computer engineering (STA)* (pp. 488–495). IEEE.
6. Junior, P. R. M., Farias, J. V. M., Cupertino, A. F., Pereira, H. A., Stopa, M. M., & de Resende, J. T. (2020). Redundancy and derating strategies for modular multilevel converter for an electric drive. *Journal of Control, Automation and Electrical Systems, 31*, 339–349.
7. El Badsi, I. N., El Badsi, B., & Masmoudi, A. (2018). RFOC and DTC strategies for reduced structure B3-inverter fed induction motor drives. In *2018 15th international multi-conference on systems, signals and devices (SSD)* (pp. 1317–1322). IEEE.

Direct Torque Control of 3-Φ Induction Motor

2.1 Introduction

In variable speed applications based on induction motor drives, there are basically two types of popular control strategies: (i) rotor flux oriented control (RFOC), and (ii) direct torque control (DTC). Both control strategies can decouple the flux and torque components, and provide good torque response in transient and steady state operation. Unlike vector control, DTC does not require any current regulators and coordinate transformation. The operating principle of DTC is simple and it controls flux and torque directly based on their instantaneous errors. In addition, DTC has less parameter sensitivity.

The earlier DTC strategy has been proposed by *Takahashi* in the middle of the 1980s [1]. Since then, many DTC strategies based on analytical foundations have been developed so far. Two major classes of DTC strategies could be distinguished: (i) strategies without controlled switching frequency [2–4], and (ii) strategies with controlled switching frequency [5–8]. Obviously, the second class offers higher performance in terms of torque ripple reduction and efficiency improvement. However, these strategies necessarily require control systems with higher CPU-frequencies in so far as their implementation schemes are more complicated than those of the first class.

Several studies investigated the possibility to associate space-vector modulation (SVM) techniques with DTC in order to control the switching frequency [5, 6]. They offer the lowest harmonic distortion associated with reduced switching losses. Further improvement could be gained through the bus-clamping SVM technique (BCSVM) [9–11].

The association of BCSVM to DTC strategies requires a high CPU-time which may compromise the control system performance. An approach to by-pass this limitation could be achieved by skipping the SVM generator and considering the bus-clamping method in the synthesis of the vector selection table of DTC strategies [12].

© The Author(s), under exclusive license to Springer Nature Switzerland AG 2025
I. Nouira and B. El Badsi, *Control Strategies of Electric Drives*, Synthesis Lectures on Power Electronics, https://doi.org/10.1007/978-3-031-81332-0_2

2.2   Space Voltage Vectors of SSTPI

It is possible to transform the stator phase voltages (v_{an}, v_{bn}, and v_{cn}) of an IM in the α-β reference frame using the power invariant transformation, such that:

$$
\begin{bmatrix} v_{\alpha s} \\ \\ v_{\beta s} \end{bmatrix} = \sqrt{\frac{2}{3}} V_{dc} \begin{bmatrix} 1 & -\frac{1}{2} & -\frac{1}{2} \\ 0 & \frac{\sqrt{3}}{2} & -\frac{\sqrt{3}}{2} \end{bmatrix} \begin{bmatrix} S_1 \\ S_2 \\ S_3 \end{bmatrix} \tag{2.1}
$$

The eight possible combinations of (S_1 S_2 S_3) define eight voltage vectors $\mathbf{V_s}$ in the α-β plane. Table 2.1 summarizes the states of the IGBTs of the SSTPI and the eight voltage vectors $\mathbf{V_s}$. As shown in Fig. 2.1, the SSTPI generates two zero states ($\mathbf{V_0}$ and $\mathbf{V_7}$) and six active states ($\mathbf{V_1}$ to $\mathbf{V_6}$). Each active vector has a $\frac{\pi}{3}$ displacement from its two adjacent vectors. These active vectors have an equal magnitude of $\sqrt{\frac{2}{3}} V_{dc}$.

Table 2.1 Switching states (S_1 S_2 S_3) of the inverter IGBTs and the voltage vectors $\mathbf{V_s}$

(S_1 S_2 S_3)	v_{an}	v_{bn}	v_{cn}	$v_{\alpha s}$	$v_{\beta s}$	$\mathbf{V_s}$
(0 0 0)	0	0	0	0	0	$\mathbf{V_0}$
(1 0 0)	$\frac{2V_{dc}}{3}$	$-\frac{V_{dc}}{3}$	$-\frac{V_{dc}}{3}$	$\sqrt{\frac{2}{3}} V_{dc}$	0	$\mathbf{V_1}$
(1 1 0)	$\frac{V_{dc}}{3}$	$\frac{V_{dc}}{3}$	$-\frac{2V_{dc}}{3}$	$\frac{V_{dc}}{\sqrt{6}}$	$\frac{V_{dc}}{\sqrt{2}}$	$\mathbf{V_2}$
(0 1 0)	$-\frac{V_{dc}}{3}$	$\frac{2V_{dc}}{3}$	$-\frac{V_{dc}}{3}$	$-\frac{V_{dc}}{\sqrt{6}}$	$\frac{V_{dc}}{\sqrt{2}}$	$\mathbf{V_3}$
(0 1 1)	$-\frac{2V_{dc}}{3}$	$\frac{V_{dc}}{3}$	$\frac{V_{dc}}{3}$	$-\sqrt{\frac{2}{3}} V_{dc}$	0	$\mathbf{V_4}$
(0 0 1)	$-\frac{V_{dc}}{3}$	$-\frac{V_{dc}}{3}$	$\frac{2V_{dc}}{3}$	$-\frac{V_{dc}}{\sqrt{6}}$	$-\frac{V_{dc}}{\sqrt{2}}$	$\mathbf{V_5}$
(1 0 1)	$\frac{V_{dc}}{3}$	$-\frac{2V_{dc}}{3}$	$\frac{V_{dc}}{3}$	$\frac{V_{dc}}{\sqrt{6}}$	$-\frac{V_{dc}}{\sqrt{2}}$	$\mathbf{V_6}$
(1 1 1)	0	0	0	0	0	$\mathbf{V_7}$

Fig. 2.1 Voltage vectors generated by a SSTPI

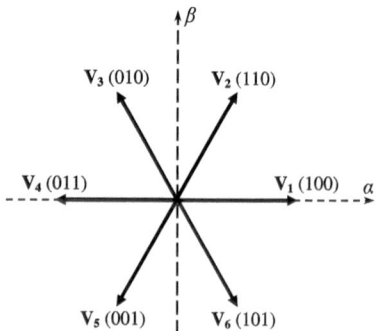

2.3   *Park* Model of IM

By applying the *Park* transform to the stator and rotor equations and excluding the zero sequence components, the following equation systems are defined:

- Voltage equations:

$$
\begin{cases}
v_{ds} = r_s i_{ds} + \dfrac{d}{dt}\phi_{ds} - \omega_s \phi_{qs} & (a) \\
v_{qs} = r_s i_{qs} + \dfrac{d}{dt}\phi_{qs} + \omega_s \phi_{ds} & (b) \\
0 = r_r i_{dr} + \dfrac{d}{dt}\phi_{dr} - \omega_r \phi_{qr} & (c) \\
0 = r_r i_{qr} + \dfrac{d}{dt}\phi_{qr} + \omega_r \phi_{dr} & (d)
\end{cases}
\tag{2.2}
$$

- Flux equations:

$$
\begin{cases}
\phi_{ds} = l_s i_{ds} + M i_{dr} & (a) \\
\phi_{qs} = l_s i_{qs} + M i_{qr} & (b) \\
\phi_{dr} = M i_{ds} + l_r i_{dr} & (c) \\
\phi_{qr} = M i_{qs} + l_r i_{qr} & (d)
\end{cases}
\tag{2.3}
$$

The substitution of the phase quantities by those of *Park* in the electromagnetic torque expression gives the following relations:

$$
\begin{cases}
T_{em} = P\ (\phi_{ds} i_{qs} - \phi_{qs} i_{ds}) & (a) \\
T_{em} = P\ (\phi_{qr} i_{dr} - \phi_{dr} i_{qr}) & (b) \\
T_{em} = \dfrac{PM}{l_r}\ (\phi_{dr} i_{qs} - \phi_{qr} i_{ds}) & (c) \\
T_{em} = \dfrac{PM}{l_s}\ (\phi_{qs} i_{dr} - \phi_{ds} i_{qr}) & (d)
\end{cases}
\tag{2.4}
$$

2.4   Principle of DTC Strategy

Considering the case where the dq-reference frame is linked to the stator (d-axis aligned with the stator a-phase), such that:

$$
\begin{cases}
\theta_s = 0 \\
\theta_r = -\theta
\end{cases}
\tag{2.5}
$$

Thus, the inverse matrix $P^{-1}(\theta_s = 0)$ used to transform the a-b-c stator variables to the d-q stator components is:

$$P^{-1} = \sqrt{\frac{2}{3}} \begin{bmatrix} 1 & -\frac{1}{2} & -\frac{1}{2} \\[2mm] 0 & \frac{\sqrt{3}}{2} & -\frac{\sqrt{3}}{2} \end{bmatrix} \tag{2.6}$$

Therefore, the d-q stator components, which are commonly named α-β stator components, are defined as follows:

$$\mathbf{X_{dq}} = \begin{bmatrix} X_d \\[2mm] X_q \end{bmatrix} = \mathbf{X_{\alpha\beta}} = \begin{bmatrix} X_\alpha \\[2mm] X_\beta \end{bmatrix} = \sqrt{\frac{2}{3}} \begin{bmatrix} 1 & -\frac{1}{2} & -\frac{1}{2} \\[2mm] 0 & \frac{\sqrt{3}}{2} & -\frac{\sqrt{3}}{2} \end{bmatrix} \begin{bmatrix} X_a \\[2mm] X_b \\[2mm] X_c \end{bmatrix} \tag{2.7}$$

2.4.1 Principle of Stator Flux Control

If a VSI configuration is maintained during a sampling period T_s, it is possible to determine the effect of such a configuration on the stator flux.

From the stator voltage Eqs. (2.2(a)) and (2.2(b)) written in a stator reference frame one can noted that:

$$\mathbf{V_s} = R_s \mathbf{I_s} + \frac{d}{dt} \mathbf{\Phi_s} \tag{2.8}$$

with:

$$\begin{cases} \mathbf{V_s} = v_{\alpha s} + j \, v_{\beta s} \\[2mm] \mathbf{I_s} = i_{\alpha s} + j \, i_{\beta s} \\[2mm] \mathbf{\Phi_s} = \phi_{\alpha s} + j \, \phi_{\beta s} \end{cases} \tag{2.9}$$

The amplitude of the voltage vector $\mathbf{V_s}$ is:

$$||\mathbf{V_s}|| = \begin{cases} \sqrt{\frac{2}{3}} V_{dc} \; ; \; \text{For active voltage vectors } (\mathbf{V_1} \text{ to } \mathbf{V_6}) \\[2mm] 0 \quad ; \; \text{For zero voltage vectors } (\mathbf{V_0} \text{ and } \mathbf{V_7}) \end{cases} \tag{2.10}$$

Neglecting the voltage drop $R_s \mathbf{I_s}$ across the stator resistance, Eq. (2.8) can be rewritten as:

$$\frac{d}{dt} \mathbf{\Phi_s} = \mathbf{V_s} \tag{2.11}$$

Therefore, for small values of T_s the variation of the stator flux vector $\boldsymbol{\Phi_s}$ between the $(k + 1)^{th}$ and the kth sampling instants can be expressed as:

$$\frac{\boldsymbol{\Phi_{s(k+1)}} - \boldsymbol{\Phi_{s(k)}}}{T_s} \simeq \mathbf{V_s} \tag{2.12}$$

with k indicates the index of the configuration applied at k^{th} sampling period, such that:

$$t_{(k+1)} - t_{(k)} = T_s \tag{2.13}$$

Then, the stator flux vector $\boldsymbol{\Phi_s}$ at $(k + 1)^{th}$ sampling period can be calculated as:

$$\boldsymbol{\Phi_{s(k+1)}} \simeq \boldsymbol{\Phi_{s(k)}} + \mathbf{V_s} T_s \tag{2.14}$$

It can be noted that the selection of a null configuration ($\mathbf{V_0}$ or $\mathbf{V_7}$) during a sampling period T_s stops the stator flux vector $\boldsymbol{\Phi_s}$ in the α-β plane. However, the application of an active configuration ($\mathbf{V_1}$ to $\mathbf{V_6}$) during a sampling period T_s moves the stator flux vector $\boldsymbol{\Phi_s}$ along the direction of the applied stator voltage vector $\mathbf{V_s}$. Therefore, the distance covered by the stator flux vector $\boldsymbol{\Phi_s}$ is proportional to the DC-bus voltage V_{dc} and to the sampling period T_s, such as:

$$||\boldsymbol{\Phi_{s(k+1)}} - \boldsymbol{\Phi_{s(k)}}|| \simeq \sqrt{\frac{2}{3}} V_{dc} T_s \tag{2.15}$$

The flux magnitude control depends on the selection of the inverter configuration to be applied to the IM. Such a control is based on the knowledge of the position of the stator flux vector $\boldsymbol{\Phi_s}$ in the α-β plane. As indicated in Fig. 2.2, the α-β plane is commonly divided into six equal sectors (I to VI), with each sector is centered by a stator voltage vector $\mathbf{V_s}$ and has an opening angle equal to $\frac{\pi}{3}$.

With an appropriate selection of the sequence of the VSI configurations the stator flux vector $\boldsymbol{\Phi_s}$ can be driven along any trajectory.

Thus, if a hysteresis regulator is used to control the stator flux, the flux hysteresis band becomes a circular region in the α-β plane, as shown in Fig. 2.3. In order to drive the flux within the hysteresis band, the VSI applies to the IM the suitable voltage vector $\mathbf{V_s}$ in each sampling period T_s. The radial component of the applied voltage vector $\mathbf{V_s}$ acts on the flux magnitude $||\boldsymbol{\Phi_s}||$, while the tangential component determines the rotation direction of the stator flux vector $\boldsymbol{\Phi_s}$.

Therefore, and in order to maintain a constant amplitude $||\boldsymbol{\Phi_s}||$ of the stator flux, a suitable choice of the applied voltage vectors $\mathbf{V_s}$ for successive sampling periods T_s, allows an operation of the IM under a constant amplitude of the stator flux. Such a flux can be obtained by moving the extremity of the stator flux vector $\boldsymbol{\Phi_s}$ in a quasi-circular path, as illustrated in Fig. 2.3.

Thus, to control the stator flux, the following rules are commonly used:

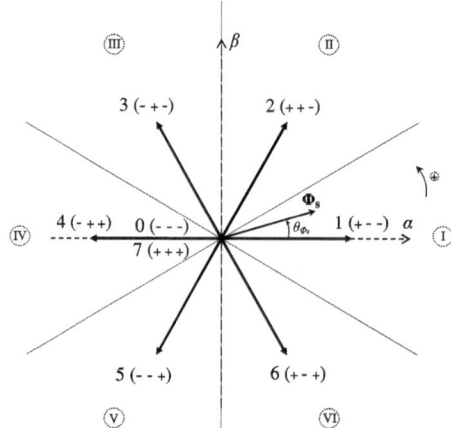

Fig. 2.2 Space voltage vectors $\mathbf{V_S}$ and stator flux vector $\mathbf{\Phi_S}$ located in the α-β frame

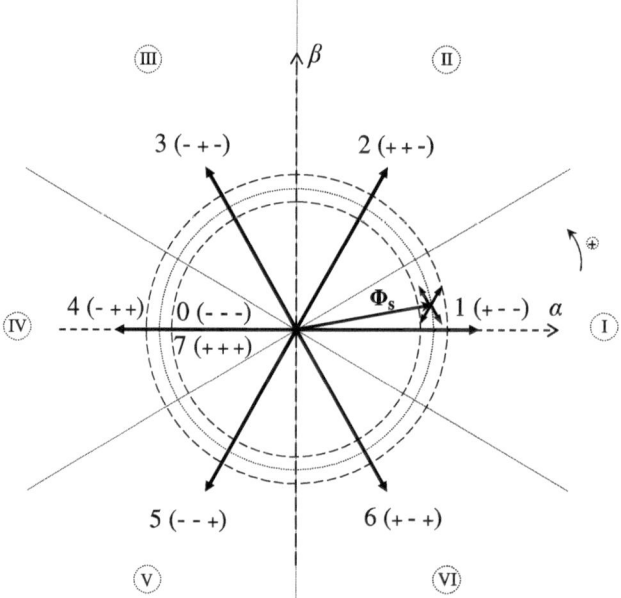

Fig. 2.3 Quasi-circular trajectory described by the extremity of the stator flux vector $\mathbf{\Phi_S}$ following the application of suitable voltage vectors $\mathbf{V_S}$

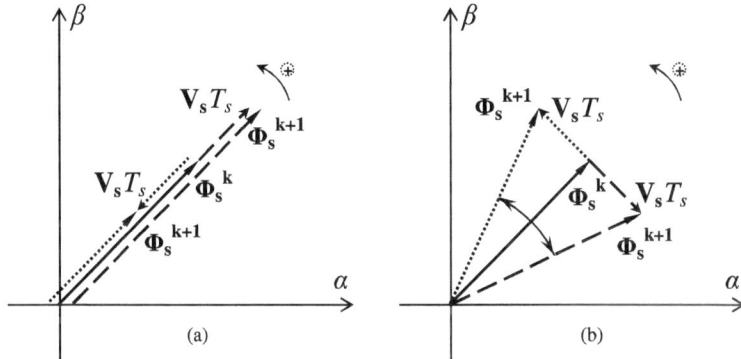

Fig. 2.4 Evolution of the stator flux vector Φ_s in the α-β plane with respect to the applied voltage vector V_s during a sampling period T_s

- The application of a voltage vector V_s parallel to the stator flux vector Φ_s with the same direction (respectively with an opposite direction) allows: (i) the increase (respectively decrease) of the stator flux amplitude, and (ii) the maintain of its phase, as illustrated in Fig. 2.4a,
- The application of a voltage vector V_s with a quadrature phase shift versus the stator flux vector Φ_s allows: (i) the holding of stator flux amplitude, and (ii) the changing of its phase, as shown in Fig. 2.4b,
- The application of a zero voltage vector V_s preserves the amplitude and phase of the stator flux vector Φ_s.

2.4.2 Principle of Electromagnetic Torque Control

From Eqs. (2.3(a)) and (2.3(b)), the d- and q-stator currents are expressed as:

$$\begin{cases} i_{ds} = \frac{1}{l_s}(\phi_{ds} - Mi_{dr}) \\ i_{qs} = \frac{1}{l_s}(\phi_{qs} - Mi_{qr}) \end{cases} \qquad (2.16)$$

The substitution of the d- and q-stator currents, given by (2.16), in the expression of the electromagnetic torque (2.4(c)) gives:

$$T_{em} = \frac{PM}{l_r}\left(\phi_{dr}\left(\frac{\phi_{qs} - Mi_{qr}}{l_s}\right) - \phi_{qr}\left(\frac{\phi_{ds} - Mi_{dr}}{l_s}\right)\right) \qquad (2.17)$$

The development of Eq. (2.17) leads to the following relation:

$$T_{em} = \frac{PM}{l_s l_r}[(\phi_{dr}\phi_{qs} - \phi_{qr}\phi_{ds}) + \underbrace{M(\phi_{qr}i_{dr} - \phi_{dr}i_{qr})]}_{T_{em}/P}$$ (2.18)

which gives:

$$T_{em}\left(1 - \frac{M^2}{l_s l_r}\right) = \frac{PM}{l_s l_r}(\phi_{dr}\phi_{qs} - \phi_{qr}\phi_{ds})$$ (2.19)

Thus, the electromagnetic torque T_{em} can be expressed, in terms of the stator and rotor flux components, as follows:

$$T_{em} = \frac{PM}{\sigma l_s l_r}(\phi_{dr}\phi_{qs} - \phi_{qr}\phi_{ds})$$ (2.20)

Equation (2.20) can be rewriting as a vectorial product of the rotor and stator flux as:

$$T_{em} = \frac{PM}{\sigma l_s l_r}||\boldsymbol{\Phi_r} \wedge \boldsymbol{\Phi_s}||$$ (2.21)

which can be developed as following:

$$T_{em} = \frac{PM}{\sigma l_s l_r}||\boldsymbol{\Phi_r}|| \, ||\boldsymbol{\Phi_s}|| \, \sin\gamma$$ (2.22)

with γ is the angle described between the rotor flux vector $\boldsymbol{\Phi_r}$ and the stator flux vector $\boldsymbol{\Phi_s}$ in counterclockwise rotation.

At any instant, the electromagnetic torque T_{em} is proportional to the stator flux amplitude $||\boldsymbol{\Phi_s}||$, the rotor flux amplitude $||\boldsymbol{\Phi_r}||$, and the sine of the angle γ between the two flux vectors.

Therefore, if the amplitude $||\boldsymbol{\Phi_s}||$ of the stator flux is maintained constant equal to its rated value, and if with such a control the rotor flux amplitude $||\boldsymbol{\Phi_r}||$ still also constant, the torque T_{em} can be controlled through the angle γ between these flux vectors.

In what follows is developed the relation between the rotor and stator flux vectors. The Eqs. (2.2(c)) and (2.2(d)) are arranged in the following vectorial relation:

$$\frac{d\boldsymbol{\Phi_r}}{dt} + j\omega_r\boldsymbol{\Phi_r} + r_r\mathbf{I_r} = 0$$ (2.23)

According to Eq. (2.3), the rotor current vector $\mathbf{I_r}$ can be expressed as following:

$$\mathbf{I_r} = \frac{1}{\sigma l_r}\left(\boldsymbol{\Phi_r} - \frac{M}{l_s}\boldsymbol{\Phi_s}\right)$$ (2.24)

Therefore, Eq. (2.23) turns to be:

$$\sigma T_r \frac{d\Phi_r}{dt} + (1 + j\sigma T_r \omega_r)\Phi_r = \frac{M}{l_s}\Phi_s \tag{2.25}$$

By considering the case of a rotor reference frame ($\frac{d\theta_r}{dt} = \omega_r = 0$) and using the *Laplace* transform, the relationship between the flux vectors is defined as:

$$\Phi_r = \frac{\frac{M}{l_s}}{1 + \sigma T_r p}\Phi_s \tag{2.26}$$

Equation (2.26) shows that the relation between the stator flux Φ_s and the rotor flux Φ_r represents a first-order low-pass filter with a time constant equal to σT_r. Thus, a variation of the stator flux Φ_s is followed by a delayed change in the rotor flux Φ_r, and therefore, if the stator flux amplitude $||\Phi_s||$ is controlled to be constant, the rotor flux amplitude $||\Phi_r||$ is also kept constant. As a conclusion, the electromagnetic torque T_{em} can be controlled through the angle γ according to Eq. (2.22).

In balanced sinusoidal steady-state condition the stator and rotor fluxes rotate at constant angular speed ω_s in a stator reference frame. The angle γ maintains a constant value, which depends on the operating conditions. In this way the produced torque T_{em} is kept constant. Starting from this condition, it is possible to analyze the effect produced by the variation of the angle γ. Taking Eqs. (2.22) and (2.26) into account leads to the following remarks:

★ if Φ_s accelerates, then γ increases and the electromagnetic torque T_{em} increases,
★ if Φ_s decelerates, then γ decreases and the electromagnetic torque T_{em} decreases.

Thus, to control the electromagnetic torque T_{em}, the following rules are usually applied:

- The application of an active voltage vector V_s parallel to the stator flux vector Φ_s maintains the flux phase, and therefore has a limited effect on the evolution of the electromagnetic torque T_{em} as illustrated in Fig. 2.4a,
- The application of a voltage vector V_s with a quadrature phase shift versus the stator flux vector Φ_s increases or decreases the electromagnetic torque T_{em}, as shown in Fig. 2.4b.

2.5   Implementation of the Classical DTC Strategy

This section deals with the implementation of the classical DTC strategy proposed by *Takahashi* [1]. Figure 2.5 shows the implementation scheme of DTC strategy for an IM, where the d-axis (α-axis) of the control frame is aligned with the stator a-phase.

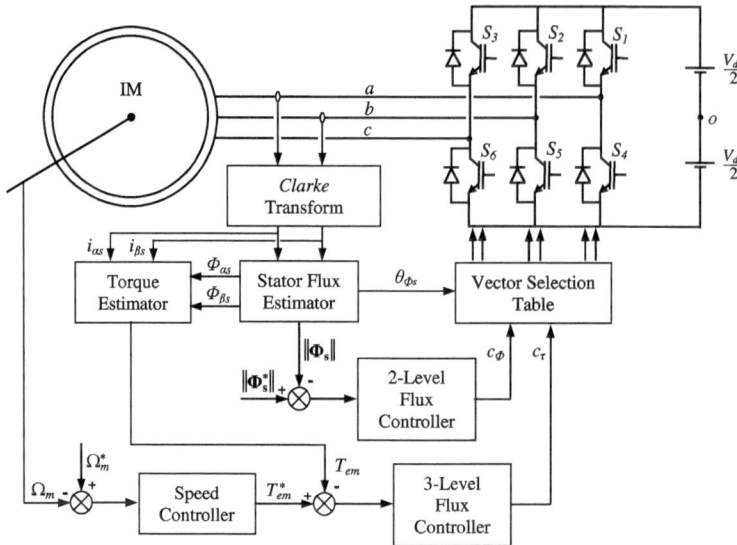

Fig. 2.5 Implementation scheme of the classical DTC strategy for SSTPI fed IM

**Control System Process**: In the following are listed the comprehensive process of the DTC control system, such as:

- Measurement of two stator currents (i_{as} and i_{bs}),
- Measurement of two stator voltages (v_{as} and v_{bs}); (alternatively the measurement of the DC-bus voltage V_{dc} in the input of the VSI, and therefore the stator voltages are determined according to the states of the inverter IGBTs),
- Determination of the *Clarke* components ($v_{\alpha s}, v_{\beta s}, i_{\alpha s}$ and $i_{\beta s}$) of the stator voltages and currents,
- Estimation of the amplitude and phase of the stator flux vector $\mathbf{\Phi_s}$ as a function of the α- and β- current and voltage components,
- Estimation of the electromagnetic torque T_{em}, which is achieved according to the α- and β- current and flux components,
- Generation of the control signal c_ϕ through the two-level hysteresis controller, which its input signal is the error between the amplitude of the reference stator flux $||\mathbf{8_s^*}||$ and the estimated one $||\mathbf{\Phi_s}||$,
- Generation of the control signal c_τ of the electromagnetic torque through a three-level hysteresis controller, which its input signal is the error between the reference torque T_{em}^* and the estimated one T_{em},
- Select the adequate combination of the IGBT states according to the angle θ_{Φ_s} of the flux vector $\mathbf{\Phi_s}$ in the α-β plane, and considering the control signals c_ϕ and c_τ.

Control Loops: The DTC control system involves three control loops, such that:

* Control loop of the mechanical speed Ω_m including a PI type controller, which generates the reference electromagnetic torque T^*_{em},
* Control loop of the electromagnetic torque T_{em} including a three-level hysteresis controller, which generates a control signal c_τ,
* Control loop of the stator flux amplitude $||\Phi_\mathbf{s}||$ including a two-level hysteresis controller, which gives a control signal c_ϕ.

Flux Estimator: From the measurement of two currents and two stator voltages, the amplitude $||\Phi_\mathbf{s}||$ and phase θ_{Φ_s} of the stator flux vector are estimated as follows:

$$||\Phi_\mathbf{s}|| = \sqrt{\phi^2_{\alpha s} + \phi^2_{\beta s}} \tag{2.27}$$

where:

$$\begin{cases} \phi_{\alpha s} = \int (v_{\alpha s} - r_s i_{\alpha s})dt \\ \phi_{\beta s} = \int (v_{\beta s} - r_s i_{\beta s})dt \end{cases} \tag{2.28}$$

and:

$$\theta_{\Phi_s} = \mathrm{tg}^{-1}\left(\frac{\phi_{\beta s}}{\phi_{\alpha s}}\right) \tag{2.29}$$

Torque Estimator: Accounting for Eq. 2.4(a), the estimated electromagnetic torque T_{em} is calculated using the α- and β-components of the stator current and stator flux vectors, such as:

$$T_{em} = P\left(\phi_{\alpha s} i_{\beta s} - \phi_{\beta s} i_{\alpha s}\right) \tag{2.30}$$

Two-Level Flux Controller: For the control loop of the stator flux, the difference between the reference flux $||8^*_\mathbf{s}||$ and the estimated flux $||\Phi_\mathbf{s}||$ from Eq. (2.27) is applied to the input of a two-level hysteresis controller. As shown in Fig. 2.6, the output of such a controller is the control signal c_ϕ, which is equal to "+1" to increase the stator flux amplitude, and equal to "−1" to decrease the stator flux amplitude.

Fig. 2.6 Two-level hysteresis controller of the stator flux

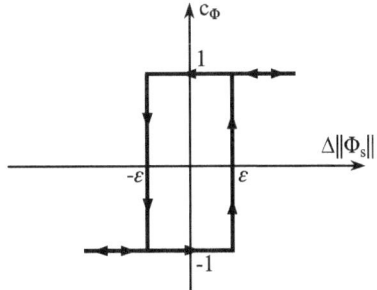

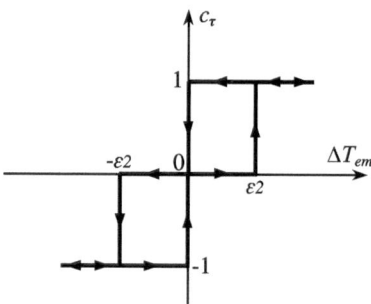

Fig. 2.7 Three-level hysteresis controller of the electromagnetic torque

Three-Level Torque Controller: For the torque control loop, the difference between the reference electromagnetic torque T_{em}^* and the estimated one T_{em} from equation (2.30) is applied to the input of a three-level hysteresis controller. As illustrated in Fig. 2.7, the output of such a controller is the control signal c_τ, which is equal to "+1" to increase the torque, "−1" to decrease the torque, and "0" to maintain the torque.

Speed Controller: The synthesis of the speed controller is carried out in the same manner as that developed in paragraph 1.4.1.3, dealing with the RFOC of IM fed by current-controlled PWM VSI.

Vector Selection Table: The vector selection table is the pre-actuator for the VSI. Such a table was established for the first time by *Takahashi*, and which allows the generation of the control signals of the inverter IGBTs.

The vector selection table includes the stator voltage vectors $\mathbf{V_s}$ (in other words the IGBT state combinations) to be generated by the VSI. The selection of the appropriate voltage vector $\mathbf{V_s}$ is based on: (i) the control signals c_ϕ and c_τ generated by the hysteresis controllers of the stator flux and the electromagnetic torque, and (ii) the angle θ_{Φ_s} giving the position of the stator flux vector $\mathbf{\Phi_s}$ in the α-β plane. The vector selection table of the classical DTC strategy is summarized in Table 2.2.

Table 2.2 Vector selection table of *Takahashi* DTC strategy

c_ϕ	+1	+1	+1	−1	−1	−1
c_τ	+1	0	−1	+1	0	−1
Sector I	V_2	V_7	V_6	V_3	V_0	V_5
Sector II	V_3	V_0	V_1	V_4	V_7	V_6
Sector III	V_4	V_7	V_2	V_5	V_0	V_1
Sector IV	V_5	V_0	V_3	V_6	V_7	V_2
Sector V	V_6	V_7	V_4	V_1	V_0	V_3
Sector VI	V_1	V_0	V_5	V_2	V_7	V_4

2.6 Bus-Clamping Based DTC Strategy

The bus-clamping based DTC strategy consists in the synthesis of two vector selection tables suitably arranged considering the bus-clamping technique. Such synthesis depends on the rotation direction of the stator flux vector Φ_s in α-β plane. The resulting DTC strategy leads to clamp each motor phase during 60° interval in every half cycle of the reference stator flux vector. Therefore, a reduction of the switching losses is gained.

The major drawback of the *Takahashi* DTC strategy is due to the systematic application of zero voltage vectors (V_0 and V_7) to maintain the electromagnetic torque (c_τ=0). The application of these voltage vectors during a sampling period T_s yields a slight decrease of the stator flux amplitude $\|\Phi_s\|$ at high speeds. However, at low speeds and especially for reduced DC-bus voltage V_{dc}, the application of zero voltage vectors leads to a high reduction of the stator flux amplitude, yielding the so-called "demagnetization phenomenon" which affects the electromagnetic torque T_{em}.

An approach to avoid this phenomenon consists in limiting the application of zero voltage vectors (V_0 and V_7) to the cases where the stator flux should be reduced which leads to the control combinations ($c_\phi = -1, c_\tau = -1$) in the case of an anti-clockwise rotation, and ($c_\phi = -1, c_\tau = +1$) in the case of a clockwise rotation. This approach represents the major rule of the bus-clamping DTC strategy (BCDTC).

2.6.1 Principle of Bus-Clamping DTC Strategy

According to Eq. (2.11) and considering the case where the stator flux vector Φ_s is aligned with the d-axis, the stator voltage V_s is aligned with the q-axis in the same direction (in the opposite direction) in the case of an anti-clockwise (in the case of a clockwise) rotation of Φ_s, as shown in Fig. 2.8.

In what follows, the study will be limited to sector I. Two cases should be distinguished depending on the rotation direction of Φ_s. Considering the case of the anti-clockwise rotation, if Φ_s is located in sector I, the corresponding reference vector V_s would then be in sub-sector II^+ which represents the second half of sector II, or in sub-sector III^-, which represents the first half of sector III.

According to the basis of SVM techniques, the approximation of the reference voltage vector V_s is achieved by the application of the two active voltage vectors $[V_2, V_3]$ and the two zero voltage vectors $[V_0, V_7]$.

Referring to Table 2.3, the application of the BCSVM strategy leads to two solutions which depend on the applied zero voltage vector in each modulation period. The first one considers the application of the zero voltage vector V_0 to clamp the third phase of the motor, named c-phase, to the DC-bus voltage low level ($LL_{V_{dc}}$). The second solution considers the application of the zero voltage vector V_7 to clamp the second phase, called b-phase, to the DC-bus voltage high level ($HL_{V_{dc}}$).

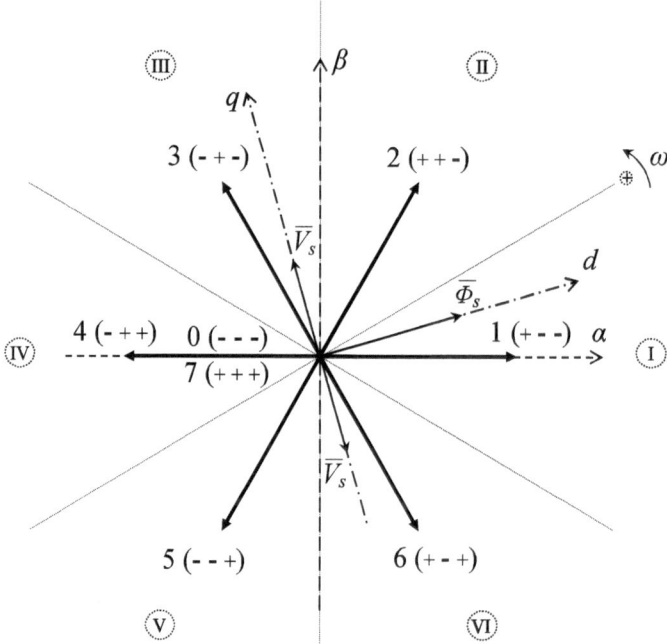

Fig. 2.8 Voltage vectors and stator reference flux in the α-β plane. (i) $\mathbf{V_s}$ is aligned with the q-axis in the same direction for an anti-clockwise rotation of $\mathbf{\Phi_s}$, (ii) $\mathbf{V_s}$ is aligned with the q-axis in the opposite direction for a clockwise rotation of $\mathbf{\Phi_s}$

2.6.2   Case of Anti-Clockwise Rotation of the Stator Flux Vector

The BCDTC strategy maintains the two active voltage vectors $\mathbf{V_2}$ and $\mathbf{V_3}$ for the corresponding control combinations ($c_\phi = +1, c_\tau = +1$) and ($c_\phi = -1, c_\tau = +1$), respectively. A zero voltage vector ($\mathbf{V_0}$ or $\mathbf{V_7}$) is applied for the control combination ($c_\phi = -1, c_\tau = -1$) in order to reduce both the stator flux amplitude $||\mathbf{\Phi_s}||$ and the electromagnetic torque T_{em}, which leads to a substitution of the active voltage vector $\mathbf{V_5}$ used in *Takahashi* DTC strategy.

Let us consider the remaining control combination ($c_\phi = +1, c_\tau = -1$) which is achieved in the *Takahashi* DTC strategy considering the active vector $\mathbf{V_6}$. However, neither vector $\mathbf{V_6}$ nor vector $\mathbf{V_5}$ could achieve the clamping of b-phase or c-phase. Therefore, the selection of the suitable active vector is restricted to $\mathbf{V_1}$ or $\mathbf{V_4}$.

Considering the case where the control combination ($c_\phi = +1, c_\tau = -1$) is achieved by the application of the active vector $\mathbf{V_1}$. This latter is suitably associated to the voltage vectors $\mathbf{V_2}$, $\mathbf{V_3}$ and $\mathbf{V_0}$ to clamp c-phase to $LL_{V_{dc}}$. The same combination achieved by the application of the active vector $\mathbf{V_4}$ allows the clamping of b-phase to $HL_{V_{dc}}$, with the association of the voltage vectors $\mathbf{V_2}$, $\mathbf{V_3}$ and $\mathbf{V_7}$, as depicted in Table 2.4.

Table 2.3 Vector selection tables of both *Takahashi* DTC and BCSVM strategies in the case of an anti-clockwise rotation

Takahashi DTC strategy		BCSVM strategy			
c_ϕ	$+1 +1 +1 -1 -1 -1$				
c_τ	$+1 \ 0 -1 +1 \ 0 -1$				
Sector of Φ_s	Sub-sectors V_s	of V_s	V_s	Clamped phase	
I	2 7 6 3 0 5	$II^+ \ III^-$	2 3 0	$c \mapsto LL_{V_{dc}}$	
			2 3 7	$b \mapsto HL_{V_{dc}}$	
II	3 0 1 4 7 6	$III^+ \ IV^-$	3 4 0	$a \mapsto LL_{V_{dc}}$	
			3 4 7	$b \mapsto HL_{V_{dc}}$	
III	4 7 2 5 0 1	$IV^+ \ V^-$	4 5 0	$a \mapsto LL_{V_{dc}}$	
			4 5 7	$c \mapsto HL_{V_{dc}}$	
IV	5 0 3 6 7 2	$V^+ \ VI^-$	5 6 0	$b \mapsto LL_{V_{dc}}$	
			5 6 7	$c \mapsto HL_{V_{dc}}$	
V	6 7 4 1 0 3	$VI^+ \ I^-$	6 1 0	$b \mapsto LL_{V_{dc}}$	
			6 1 7	$a \mapsto HL_{V_{dc}}$	
VI	1 0 5 2 7 4	$I^+ \ II^-$	1 2 0	$c \mapsto LL_{V_{dc}}$	
			1 2 7	$a \mapsto HL_{V_{dc}}$	

Figure 2.9a shows that the application of vector V_1 leads to:

- an increase of both flux and torque, corresponding to $(c_\phi = +1, c_\tau = +1)$, in the first half of sector I, named sub-sector I^-,
- an increase of the flux and a decrease of the torque, corresponding to $(c_\phi = +1, c_\tau = -1)$, in the second half of sector I, named sub-sector I^+.

Thus, the control combination $(c_\phi = +1, c_\tau = -1)$ is achieved by the application of the voltage vector V_1 only in sub-sector I^+.

The same approach has been adopted in the case of the application of the voltage vector V_4, which yields:

- a decrease of both flux and torque, corresponding to $(c_\phi = -1, c_\tau = -1)$, in sub-sector I^-,
- a decrease of the flux and an increase of the torque, corresponding to $(c_\phi = -1, c_\tau = +1)$, in sub-sector I^+.

Therefore, the control combination $(c_\phi = +1, c_\tau = -1)$ could not be achieved by the application of the active voltage vector V_4. As a result and considering sector I, the only possible solution is the clamping of c-phase through the application of the active voltage vector V_1.

Table 2.4 Preliminary vector selection table of the BCDTC strategy in the case of an anti-clockwise rotation

c_ϕ	+1	+1	−1	−1	Clamped
c_τ	+1	−1	+1	−1	phase
Sector I	2	1	3	0	$c \mapsto LL_{V_{dc}}$
	2	4	3	7	$b \mapsto HL_{V_{dc}}$
Sector II	3	5	4	0	$a \mapsto LL_{V_{dc}}$
	3	2	4	7	$b \mapsto HL_{V_{dc}}$
Sector III	4	3	5	0	$a \mapsto LL_{V_{dc}}$
	4	6	5	7	$c \mapsto HL_{V_{dc}}$
Sector IV	5	1	6	0	$b \mapsto LL_{V_{dc}}$
	5	4	6	7	$c \mapsto HL_{V_{dc}}$
Sector V	6	5	1	0	$b \mapsto LL_{V_{dc}}$
	6	2	1	7	$a \mapsto HL_{V_{dc}}$
Sector VI	1	3	2	0	$c \mapsto LL_{V_{dc}}$
	1	6	2	7	$a \mapsto HL_{V_{dc}}$

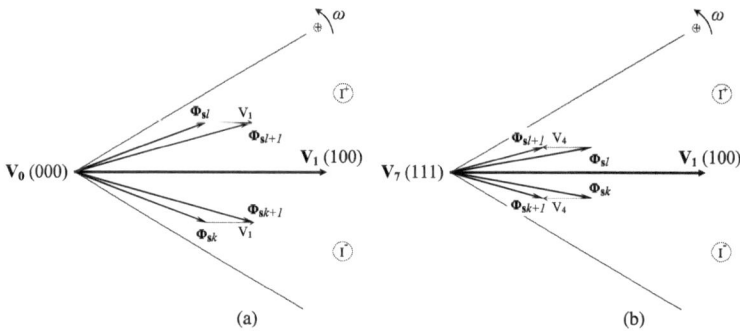

(a)   (b)

Fig. 2.9 Evolution of the stator flux vector Φ_s in sub-sectors I^- and I^+ following the application of the active voltage vectors V_1 and V_4

This said, one should solve the problem associated to the application of vector V_1 when Φ_s is located in sub-sector I^-.

The approach adopted in the BCDTC consists in applying consecutively the zero voltage vector V_0 and the active voltage vector V_1 to increase the stator flux and to decrease the electromagnetic torque. Figure 2.10a shows the evolutions of the stator flux and the electromagnetic torque in sub-sector I^- following the application of the sequence "$V_0;V_1;V_0;V_1$" during four consecutive sampling periods. For the sake of the simplicity of the vector selection table and therefore the reduction of the CPU time, a sequence of voltage vectors "$V_0;V_1;V_0;V_1$" has been adopted for both sub-sectors of sector I.

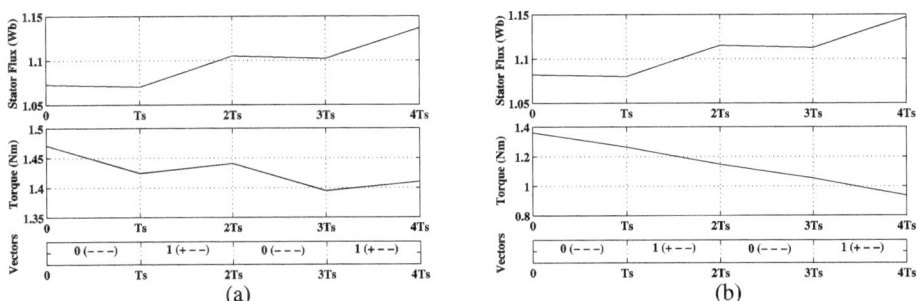

Fig. 2.10 Evolutions of the stator flux and the electromagnetic torque over four sampling periods following the application of the sequence: "$V_0;V_1;V_0;V_1$" for an anti-clockwise rotation of Φ_s in sector I. Legend: **a** case of sub-sector I$^-$, **b** case of sub-sector I$^+$.

Figure 2.11 summarizes the control combinations in the case of an anti-clockwise rotation of Φ_s in sector I of the α-β plane.

The resulting vector selection table, in the case of an anti-clockwise rotation of the stator flux vector Φ_s, is provided in Table 2.5.

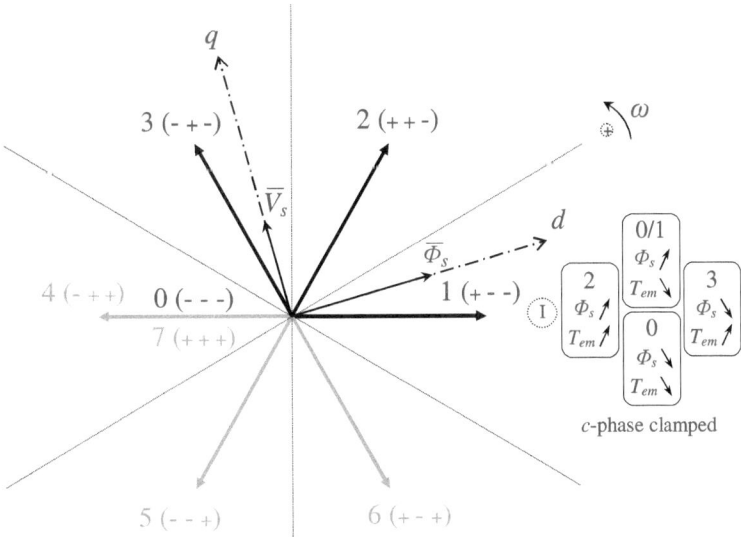

Fig. 2.11 Control combinations and the corresponding voltage vectors in the case of an anti-clockwise rotation of Φ_s in sector I

Table 2.5 Vector selection table of the BCDTC strategy for an anti-clockwise rotation

c_ϕ	+1	+1	−1	−1	Clamped
c_τ	+1	−1	+1	−1	phase
Sector I	2	0/1	3	0	$c \mapsto LL_{V_{dc}}$
Sector II	3	7/2	4	7	$b \mapsto HL_{V_{dc}}$
Sector III	4	0/3	5	0	$a \mapsto LL_{V_{dc}}$
Sector IV	5	7/4	6	7	$c \mapsto HL_{V_{dc}}$
Sector V	6	0/5	1	0	$b \mapsto LL_{V_{dc}}$
Sector VI	1	7/6	2	7	$a \mapsto HL_{V_{dc}}$

2.6.3   Case Study: Simulation with $Matlab-Simulink$ and dSPACE 1104 Experimental Validation of a DTC Program

The ratings and parameters of the three phase IM, considered in both simulation and experiments, are provided in Tables 2.6 and 2.7, respectively (Fig 2.12).

Simulation works have been carried out to investigate the performance of the classical DTC (CDTC) and bus-clamping (BCDTC) strategies. Beyond the time-varying variables, the comparison is extended to the criteria defined as follows:

1. Average switching frequency f_s: The first criterion is the average switching frequency of the three top inverter IGBTs, which is defined by:

$$f_s = \frac{\sum_{k=1}^{3} N_{c_k}}{3T_o} \tag{2.31}$$

N_{c_k}: number of commutations of the kth IGBT, during the stator current period T_o.

Table 2.6 Induction motor ratings

Power	0.37 kW	Efficiency	77%
Star	400 V/1 A	Delta	230 V/1.7A
Torque	2.56 N.m	Stator flux (rms)	0.64 Wb
Speed	1380 rpm	Frequency	50 Hz

Table 2.7 Induction motor parameters

$r_s = 24.6\,\Omega$	$l_s = 0.984\,H$	$M = 0.914\,H$	$J = 0.0025\,kg.m^2$
$r_r = 17.9\,\Omega$	$l_r = 0.984\,H$	$P = 2$	$f = 0.006\,N.m.s$

Fig. 2.12 Simulink model of the DTC strategy for an IM

2. Average total harmonic distortion THD: The second criterion is the average THD of the three stator currents, which is expressed as:

$$\text{THD} = \frac{1}{3} \sum_{k=1}^{3} \frac{\sqrt{\sum_{n \neq 1} I_{s_{kn}}^2}}{I_{s_{k1}}} \tag{2.32}$$

$I_{s_{k1}}$ and $I_{s_{kn}}$: rms values of the fundamental and n^{th} harmonic currents, respectively.

3. Average switching loss factor SLF: The third criterion is the average switching loss factor, which is defined as follows:

$$\text{SLF} = \frac{1}{3} \sum_{k=1}^{3} \sum_{j=1}^{N_{c_k}} i_{s_k \cdot j}^2 \tag{2.33}$$

where $i_{s_k, j}$ denotes the instantaneous value of the stator current of the phase connected to the kth IGBT at the jth switching instant. At a constant supply voltage, the switching energy loss in a power switch is proportional to the product of the switched current and the switching time (turn-on and turn off). The switching time, however, depends of the magnitude of the switched current. For the sake of generality, it has been assumed that the switching time is directly proportional to the current magnitude, which has led to expression (2.33).

2.6.4 Comparative Analysis of Simulated and Experimental Performance

Simulation Results: The sampling period T_s is equal to 100μs. The reference stator flux is kept constant equal to $\sqrt{3}$ times its rated value. A band width of the flux controller is equal to ± 0.02Wb which represents $\pm 1.8\%$ of the reference stator flux. The one of torque controller is equal to ± 0.04Nm which represents $\pm 1.6\%$ of the rated torque.

Figures 2.13 and 2.14 shows some simulation results of the induction motor drive at steady-state operation characterized by a reference mechanical speed $\Omega_m = +80$ rad/s, a constant load torque $T_l = 1$N.m and a DC-bus voltage $V_{dc} = 450$ V. The waveforms presented in the left-side (subscript "1") and those in the right-side (subscript "2") of Figs. 2.13 and 2.14 represent the results yielded by the CDTC and BCDTC strategies, respectively.

Referring to Fig. 2.13b2, c2, and (d2), one can notice that in each operating cycle, the BCDTC strategy allows the clamping of the top-IGBT in "ON" and in "OFF" states. Each clamping state takes place during a 60° duration, which represents a crucial efficiency benefit as far as the inverter switching losses are reduced. Such a benefit is missed in the CDTC strategy.

Furthermore, BCDTC method leads to a stator voltage v_{as} waveform quiet similar to the ones obtained under SVM techniques. The harmonic distortion of such stator voltage could be less than that of CDTC. The average THD of the three currents is about 14.1% in the case of CDTC, and about 12.7% under the BCDTC.

Moreover, it is clear that while the flux presents almost the same ripple amplitude under both DTC strategies, the one of the torque is lower under BCDTC strategy.

Now let us consider the application of the above-described comparison criteria over a range of the stator frequencies varying from 3 to 50 Hz considering two levels of the DC-bus voltage, such that: $V_{dc} = 450$V (subscript "1") and $V_{dc} = 650$ V (subscript "2"). The corresponding simulation results are illustrated in Fig. 2.15.

One can notice that for low stator frequencies, BCDTC strategy leads to a lower switching frequency f_s than the one yielded by CDTC strategy. The gap between the switching frequencies of both strategies turns to be higher with the increase of V_{dc}. However, for high stator frequencies, both strategies yield similar average values of f_s.

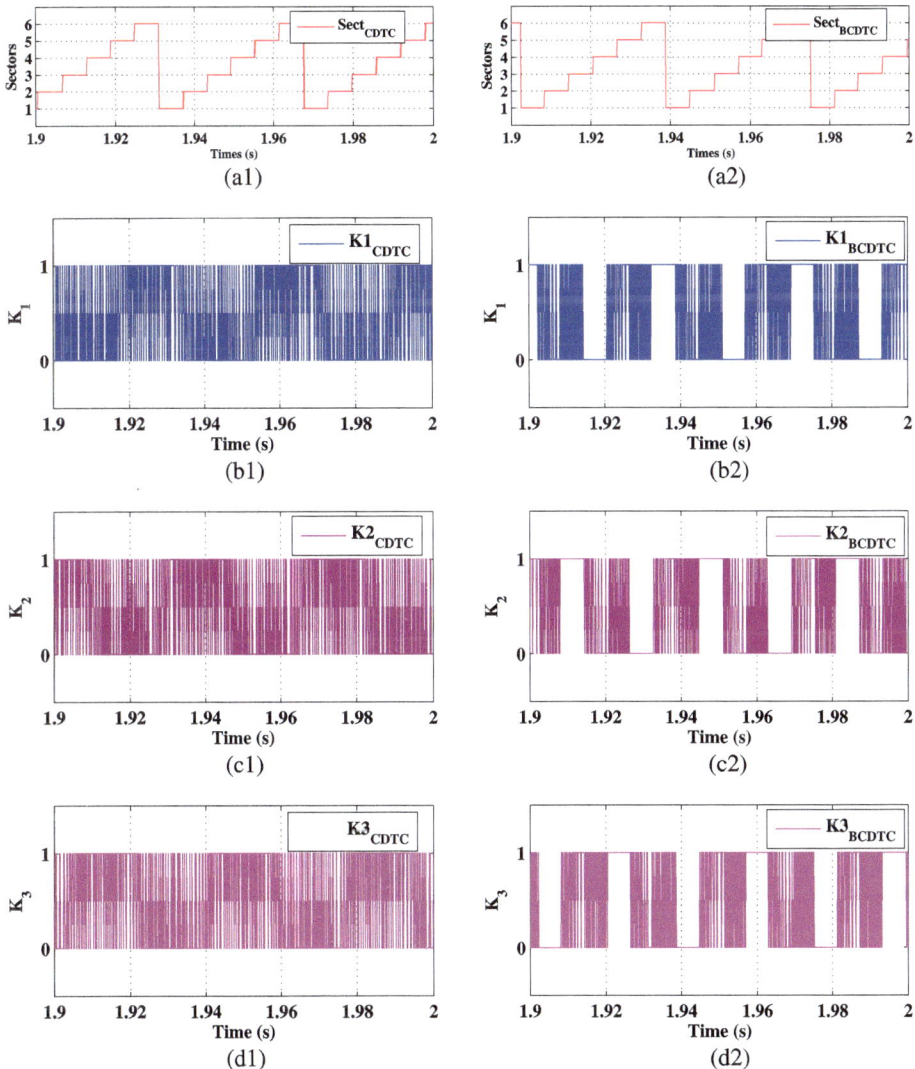

Fig. 2.13 Simulation results considering CDTC (subscript "1") and BCDTC (subscript "2") for a speed $\Omega_m = +80$ rad/s and a load torque $T_l = 1$ N.m. **Legend a**: sectors, **b, c,** and **d**: states of the three top IGBTs

Referring to the THD curves, one can notice that in the whole stator frequency range, the BCDTC strategy exhibits a lowest distortion of the stator currents. Once more, the gap between the two THD curves turns to be higher with the increase of V_{dc}.

Concerning the profiles of the switching loss criterion SLF, one can notice that they are in full harmony with those of the switching frequency f_s.

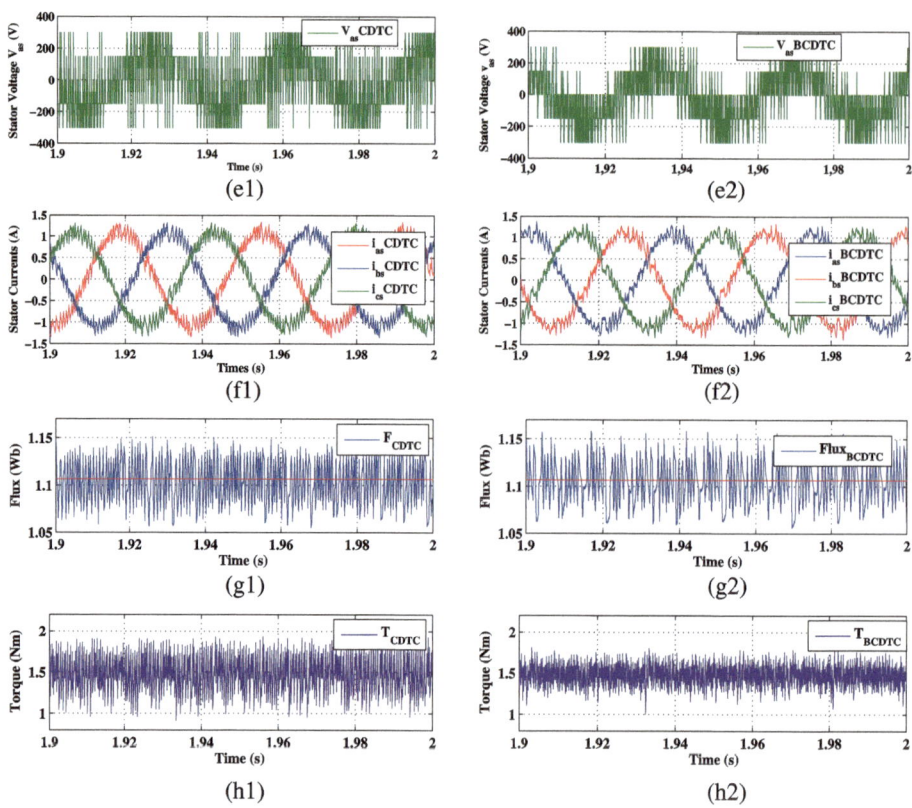

Fig. 2.14 Simulation results considering CDTC (subscript "1") and BCDTC (subscript "2") for a speed $\Omega_m = +80$rad/s and a load torque $T_l = 1$N.m. **Legend e**: stator voltage v_{as}, **f**: stator currents, **g**: stator flux, and **h**: electromagnetic torque

Experimental Validation: Both CDTC and BCDTC strategies have been implemented in a test bench built around a TMS320F240 DSP-based digital controller. A schematic block diagram of the experimental platform is shown in Fig. 2.16.

The sampling period, the reference stator flux, and the band widths of the controllers have been kept the same as in simulation. Subscripts "1" and "2" are assigned to the experimental results yielded by the CDTC strategy and the BCDTC one, respectively.

Figure 2.17 illustrates the experimental results under no-load operation considering a speed of +80rad/s (anti-clockwise rotation) and a DC-bus voltage of 400 V. Figure 2.17a1 and a2 show the control signal of the top-IGBT of the inverter leg connected to a-phase and the corresponding stator voltage v_{as}. One can confirm that the a-phase is clamped twice per cycle to high and low levels of the DC-bus voltage, following the implementation of the BCDTC strategy.

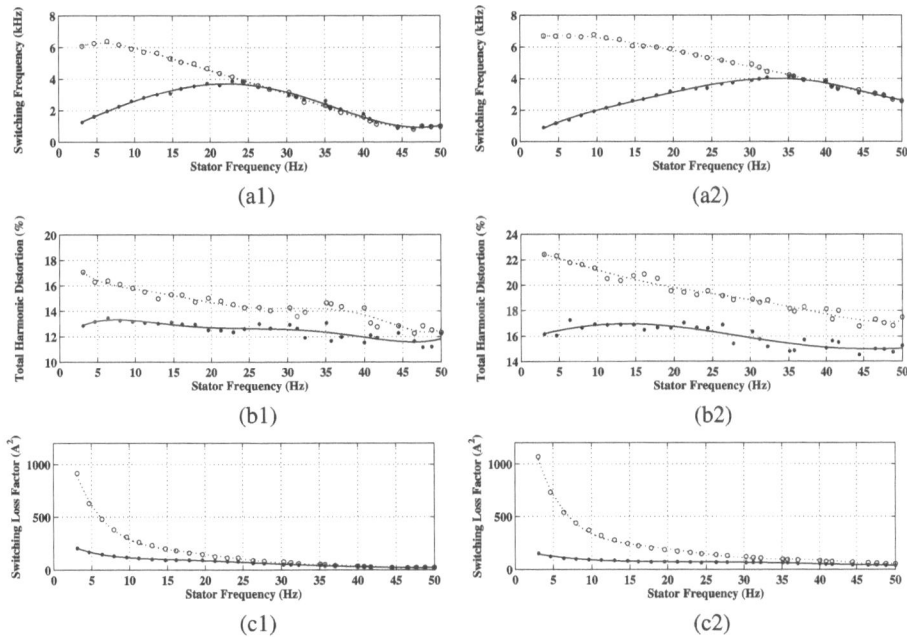

Fig. 2.15 Simulation results considering CDTC (dashed line) and BCDTC (continuous line) for a load torque $T_l = 1$N.m. **Legend 1 a**: average switching frequency (f_s) of IGBTs, **b**: average total harmonic distortion (THD) of stator currents, and **c**: average switching loss factor (SLF). **Legend 2** subscript (1): DC-voltage $V_{dc} = 450$V, and subscript (2): DC-voltage $V_{dc} = 650$V

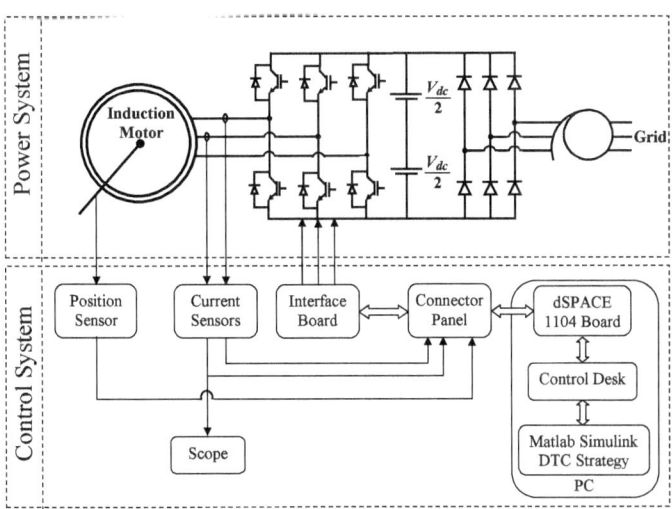

Fig. 2.16 Schematic block diagram of the experimental platform

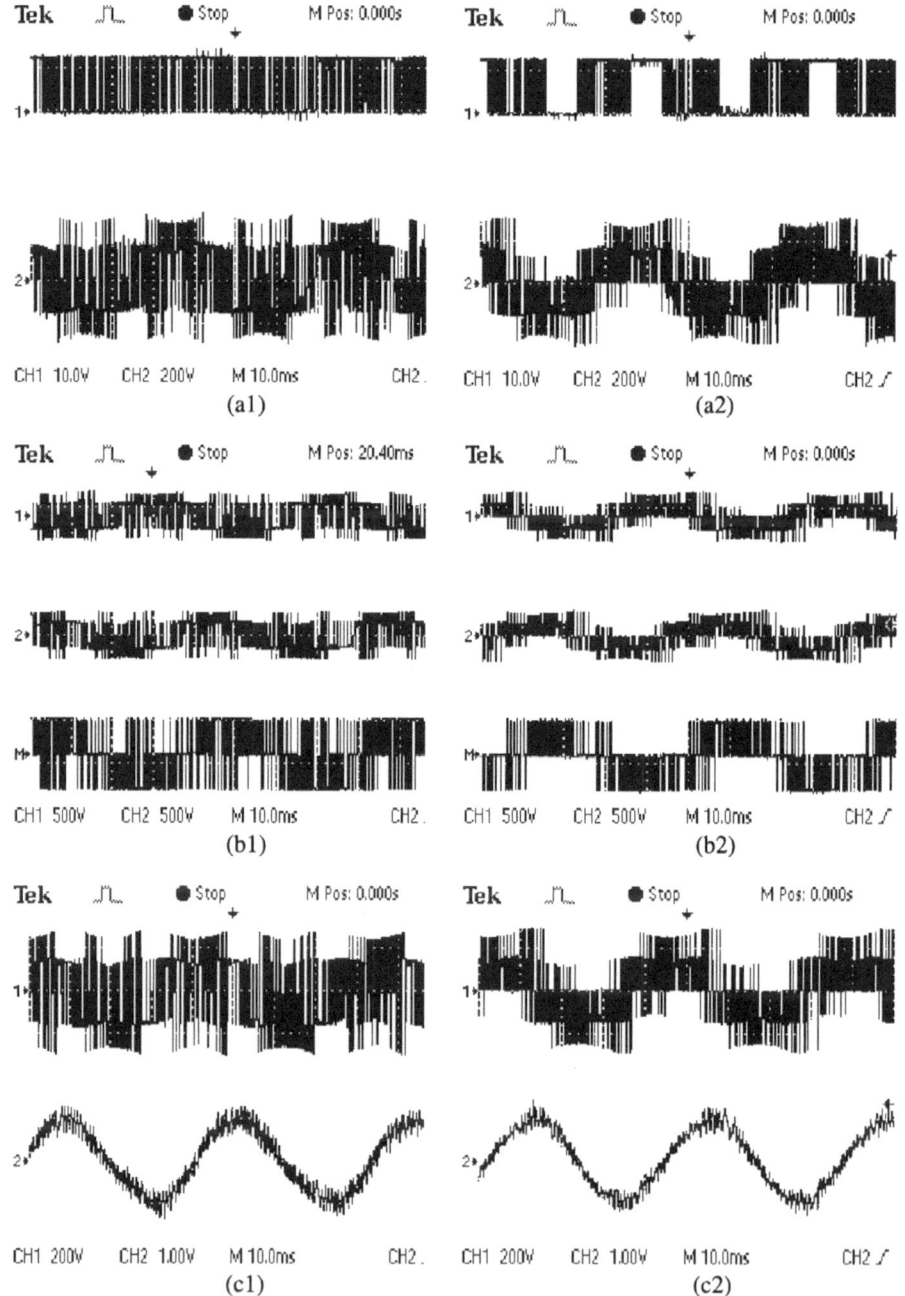

Fig. 2.17 Experimental results considering CDTC (subscript "1") and BCDTC (subscript "2") at no-load operation. **Legend a**: control signal of the first top IGBT and stator voltage v_{as} corresponding to the a-phase, **b**: stator voltages v_{as}, v_{bs} and line-to-line voltage U_{ab}, **c**: stator voltage v_{as} and stator current i_{as} (1A/div)

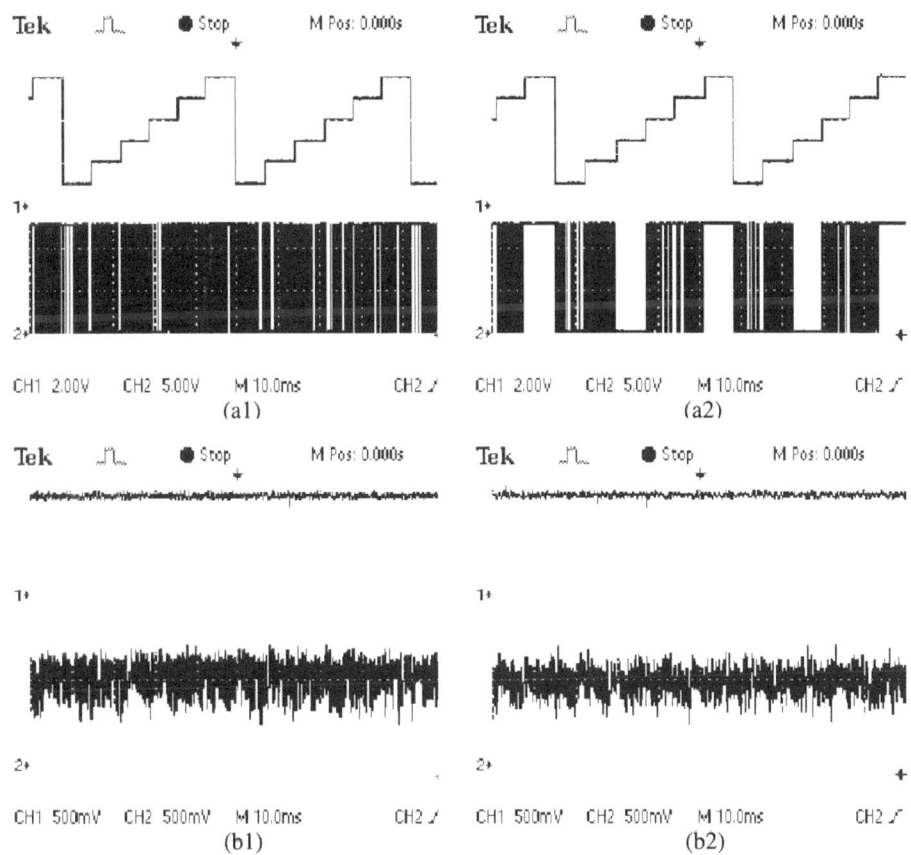

Fig. 2.18 Experimental results considering CDTC (subscript "1") and BCDTC (subscript "2") at a load torque $|T_l| = 1$ Nm. **Legend a**: sectors (2 sectors/div) and control signal of the first top IGBT, **b**: stator flux (0.5 Wb/div) and electromagnetic torque (0.5 N.m/div)

Figure 2.18 treats the operation under a constant load torque of +1N.m with a reference speed of +80rad/s (Figs. (a) and (b)) and Fig. 2.19 a constant load torque of −1N.m with a reference speed of -80rad/s (Figs. (c) and (d)), under a DC-bus voltage equal to 400V.

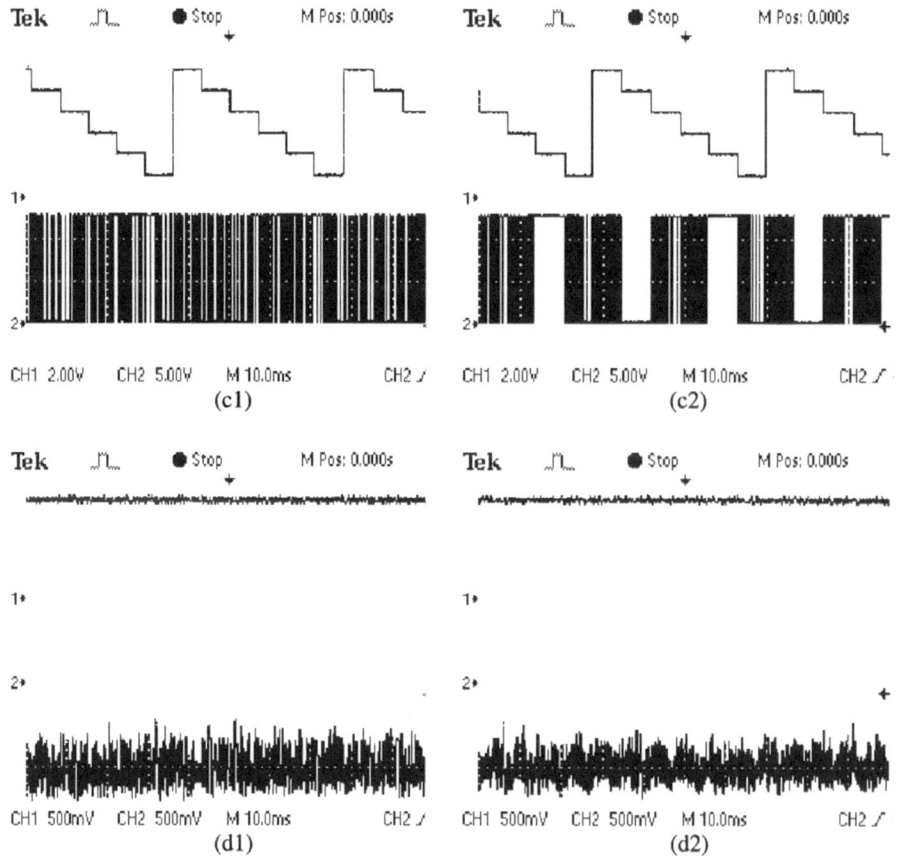

Fig. 2.19 Experimental results considering CDTC (subscript "1") and BCDTC (subscript "2") at a load torque $|T_l| = -1\,\mathrm{Nm}$ with a reference speed of $-80\,\mathrm{rad/s}$. **Legend c**: sectors (2 sectors/div) and control signal of the first top IGBT, **d**: stator flux (0.5 Wb/div) and electromagnetic torque (0.5 N.m/div)

2.7   Conclusion

In this chapter, we have explored the principles and implementation of Direct Torque Control (DTC) for induction motors. DTC stands out as a powerful control strategy that offers superior performance in terms of dynamic response and torque regulation compared to traditional methods. By leveraging the Park transformation, we can effectively model and control the induction motor in a rotating reference frame, simplifying the complexities associated with three-phase systems.

The discussion of space voltage vectors in the context of the Switch-Structured Three-Phase Inverter (SSTPI) provided valuable insights into how voltage vector selection influences motor performance. Furthermore, we examined the core principles of stator flux and electromagnetic torque control, which are fundamental to achieving precise and rapid motor control.

The implementation of the classical DTC strategy demonstrates its practical viability, showcasing the necessary algorithms and hardware configurations that contribute to its success. Additionally, the exploration of bus-clamping based DTC strategies highlights innovative approaches to enhancing control performance, particularly in situations requiring anti-clockwise rotation of the stator flux vector.

Overall, DTC represents a significant advancement in induction motor control, providing robust solutions for various applications, from industrial drives to renewable energy systems. As we continue to refine these strategies and integrate them with advanced simulation tools like Matlab/Simulink and practical validation techniques such as dSPACE 1104, we can expect further improvements in efficiency and reliability, paving the way for more sophisticated and responsive motor control systems.

References

1. Takahashi, I., & Noguchi, T. (1986). A new quick-response and high-efficiency control strategy of an induction motor. *IEEE Trans. Ind. Appl., 22*(5), 820–827.
2. Khoucha, F., Lagoun, S. M., Marouani, K., Kheloui, A., & Benbouzid, M. E. H. (2010). Hybrid cascaded h-bridge multilevel-inverter induction-motor-drive direct torque control for automotive applications. *IEEE Trans. Ind. Electron., 57*(3), 892–899.
3. Patel, C., Rajeevan, P. P., Dey, A., Ramchand, R., Gopakumar, K., & Kazmierkowski, M. P. (2012). Fast direct torque control of an open-end induction motor drive using 12-sided polygonal voltage space vectors. *IEEE Trans. Power Electron., 27*(1), 400–410.
4. El Badsi, B., Bouzidi, B., & Masmoudi, A. (2013). DTC scheme for a four-switch inverter fed induction motor emulating the six-switch inverter operation. *IEEE Trans. Power Electron., 28*(7), 3528–3538.
5. Lee, K. B., & Blaabjerg, F. (2007). An improved DTC-SVM method for sensorless matrix converter drives using an overmodulation strategy and a simple nonlinearity compensation. *IEEE Trans. Ind. Electron., 54*(6), 3155–3166.
6. Shyu, K. K., Lin, J. K., Pham, V. T., Yang, M. J., & Wang, T. W. (2010). Global minimum torque ripple design for direct torque control of induction motor drives. *IEEE Trans. Ind. Electron., 57*(9), 3148–3156.
7. Jidin, A., Idris, N. R. N., Yatim, A. H. M., Sutikno, T., & Elbuluk, M. E. (2011). Simple dynamic overmodulation strategy for fast torque control in DTC of induction machines with constant-switching-frequency controller. *IEEE Trans. Ind. Appl., 47*(5), 2283–2291.
8. Zhang, Y., & Zhu, J. (2011). A novel duty cycle control strategy to reduce both torque and flux ripples for DTC of permanent magnet synchronous motor drives with switching frequency reduction. *IEEE Trans. Power Electron., 26*(10), 3055–3067.
9. Narayanan, G., Krishnamurthy, H. K., Zhao, D., & Ayyanar, R. (2006). Advanced bus-clamping PWM techniques based on space vector approach. *IEEE Trans. Power Electron., 21*(4), 974–984.

10. Narayanan, G., Zhao, D., Krishnamurthy, H. K., Ayyanar, R., & Ranganathan, V. T. (2008). Space vector based hybrid PWM techniques for reduced current ripple. *IEEE Trans. Ind. Electron., 55*(4), 1614–1626.
11. Basu, K., Prasad, J. S. S., Narayanan, G., Krishnamurthy, H. K., & Ayyanar, R. (2010). Reduction of torque ripple in induction motor drives using an advanced hybrid PWM technique. *IEEE Trans. Ind. Electron., 57*(6), 2085–2091.
12. El Badsi, B., Bouzidi, B., & Masmoudi, A. (2013). Bus-clamping-based DTC: an attempt to reduce harmonic distortion and switching losses. *IEEE Trans. Ind. Electron., 60*(3), 873–884.

Direct Power Control of Grid-Connected DC/AC Converters

3

3.1 Introduction

The rapid growth of AC adjustable speed drives in industry aggravates the problem of harmonic pollution of the power system caused by the commonly used diode rectifiers. The use of PWM rectifiers constitutes the best solution. These rectifiers have an additional advantage of the bi-directional power flow. Control techniques for PWM rectifiers can generally be classified as *voltage based* and *virtual-flux based*. Generally, four types of these techniques can be distinguished [1, 4]:

- Voltage oriented control (VOC),
- Voltage-based direct power control (V-DPC),
- Virtual-flux oriented control (VFOC),
- Virtual-flux-based direct power control (VF-DPC).

Various control strategies have been proposed in recent works on this type PWM converter. Although these control strategies can achieve the same main goals, such as the high power factor and near-sinusoidal current waveforms, their principles differ.

The VOC, which guarantees high dynamics and static performance via an internal current control loops, has become very popular and has constantly been developed and improved. Consequently, the final configuration and performance of the VOC system largely depends on the quality of the applied current control strategy [5].

Another control strategy called Direct Power Control (DPC) is based on the instantaneous active and reactive power control loops. In DPC there are no internal current control loops and no PWM modulator block, because the converter switching states are selected by a switching table based on the instantaneous errors between the commanded and estimated values of active and reactive power. Therefore, the key point of the DPC implementation is a correct and fast estimation of the active and reactive line power [7].

© The Author(s), under exclusive license to Springer Nature Switzerland AG 2025
I. Nouira and B. El Badsi, *Control Strategies of Electric Drives*, Synthesis Lectures on Power Electronics, https://doi.org/10.1007/978-3-031-81332-0_3

3.2   Rectifier Topologies

A voltage source PWM inverter with diode rectifier is one of the most common power configuration used in modern variable speed AC drives (Fig. 3.1). An uncontrolled diode rectifier has the advantage of being simple, robust and low cost [6]. However, it allows only undirectional power flow. Therefore, energy returned from the motor must be dissipated on power resistor controlled by chopper connected across the DC link. The diode input circuit also results in lower power factor and high level of harmonic input currents.

Equation (3.1) can be used to determine the order and magnitude of the harmonic currents drawn by a six-pulse diode rectifier:

$$\begin{cases} \frac{I_h}{I_1} = \frac{1}{h} \\ h = 6k \pm 1; \quad k = 1, 2, 3, \dots \end{cases} \tag{3.1}$$

Harmonic orders as multiples of the fundamental frequency: 5, 7, 11, 13th ..., with a 50 Hz fundamental, corresponds to 250, 350, 550 and 650 Hz, respectively.

The magnitude of the harmonics in per unit of the fundamental is the reciprocal of the harmonic order: 20% for the 5th, 14.3% for the 7th. Equation (3.1) is calculated from the *Fourier* series for ideal square wave current.

Besides of six-pulse bridge rectifier a few other rectifier topologies are known. Figure 3.2 presents the main circuit of the PWM rectifier, which is the most popular topology used in adjustable speed drives. This universal topology has the advantage of using a low-cost three-phase module with a bi-directional energy flow capability.

In a AC/DC/AC converter (Fig. 3.3), the AC power is first transformed into DC thanks to three-phase PWM rectifier. It provides unit power factor and low current harmonic content. The converter connected to the DC-bus provides further desired conversion for the loads,

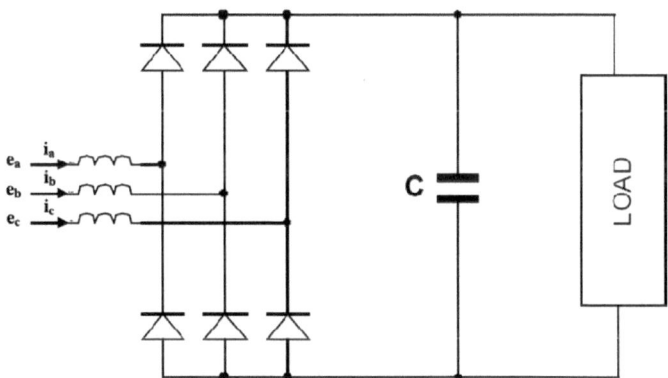

Fig. 3.1 Diode rectifier

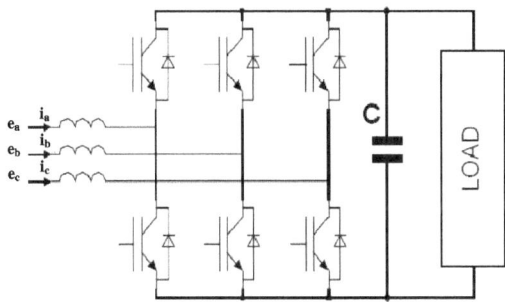

Fig. 3.2 Three-phase PWM reversible rectifier

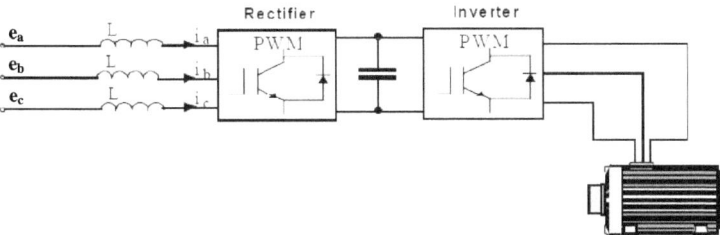

Fig. 3.3 AC/DC/AC converter

such as adjustable speed drives for induction motors (IM) or permanent magnet synchronous motors (PMSM).

3.3   Modeling of the Three-Phase PWM Rectifier

3.3.1   Operation of the PWM Rectifier

The topology of the three-phase bidirectional PWM converter is shown in Fig. 3.4. It consists of three legs with IGBTs or, in case of high power GTOs. The converter is connected to the three phase AC power supply via a smoothing inductance L and internal resistance R for each phase. The inductance L acts as a line filter for smoothing the line current with minimum ripples. The IGBTs are used as the converter bidirectional switches.

By assuming a balanced three-phase system, the voltage equation of the PWM controlled rectifier can be described as following:

$$\begin{bmatrix} v_{an} \\ v_{bn} \\ v_{cn} \end{bmatrix} = \begin{bmatrix} e_a \\ e_b \\ e_c \end{bmatrix} - R \begin{bmatrix} i_a \\ i_b \\ i_c \end{bmatrix} - L \frac{d}{dt} \begin{bmatrix} i_a \\ i_b \\ i_c \end{bmatrix} \tag{3.2}$$

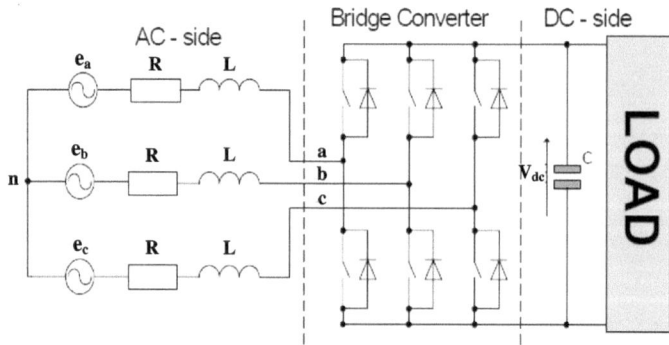

Fig. 3.4 Main circuit of three-phase PWM rectifier for bi-directional power flow

where e_{abc} is the three-phase voltage supply, i_{abc} is the three-phase line current, and $v_{abc,n}$ is the three-phase converter pole voltage, which depends on the power switch states and DC voltage level.

If the line resistance R is negligible, the line current i_{abc} is controlled by the voltage drop v_L across the inductance L interconnecting two voltage sources (line e_{abc} and converter $v_{abc,n}$). It means that the inductance voltage v_L is equal to the difference between the line voltage e_{abc} and the converter voltage $v_{abc,n}$.

With the control of the amplitude and phase angle of converter voltage $v_{abc,n}$, one can control the three-phase line current i_{abc} and, therefore, the control of the active power P conducted through the PWM converter. The reactive power Q can be controlled independently with shift of fundamental current i_{abc} with respect to line voltage e_{abc}.

Figure 3.5 presents general phasor diagram and both rectification and regenerating phasor diagrams when unity power factor is required. The PWM converter voltage can be represented with eight possible switching states (six-active and two-zero).

3.3.2   Mathematical Description of the PWM Rectifier

The basic relationship between vectors of the PWM rectifier is presented in Fig. 3.6.

Line voltages and currents The three-phase line voltages and the fundamental line currents are:

$$\begin{cases} e_a = E_m \cos(\omega t) \\ e_b = E_m \cos(\omega t + \frac{2\pi}{3}) \\ e_c = E_m \cos(\omega t + \frac{4\pi}{3}) \end{cases} \tag{3.3}$$

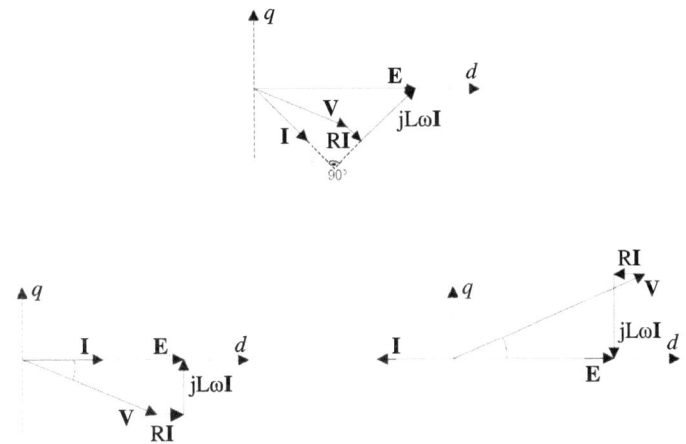

Fig. 3.5 Phasor diagram for the PWM rectifier: **a** general phasor diagram, **b** rectification at unity power factor, and **c** inversion at unity power factor

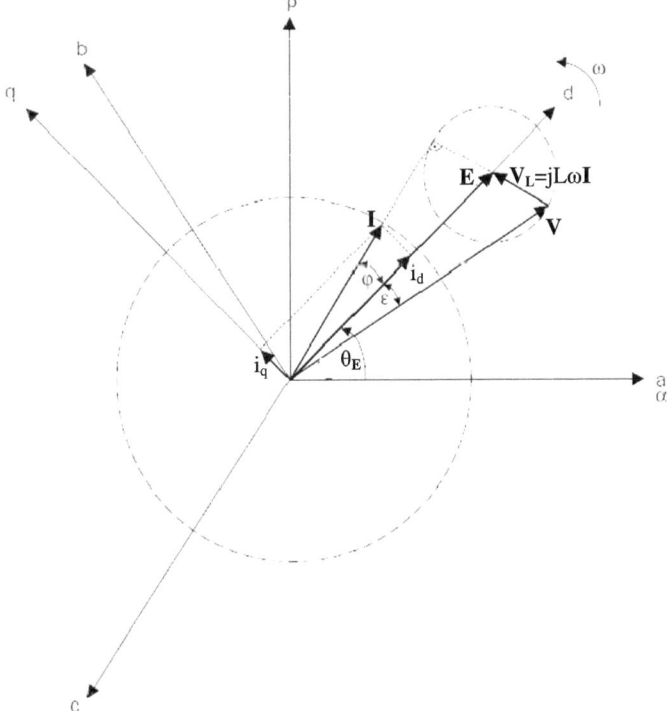

Fig. 3.6 Relationship between vectors in PWM rectifier

$$\begin{cases} i_a = I_m \cos(\omega t + \varphi) \\ i_b = I_m \cos(\omega t + \frac{2\pi}{3} + \varphi) \\ i_c = I_m \cos(\omega t + \frac{4\pi}{3} + \varphi) \end{cases} \tag{3.4}$$

where E_m (I_m) and ω are amplitude of the line voltage (current) and angular frequency, respectively.

One can transform Eq. (3.3) to α-β system thanks to *Clarke* transform, and the input voltage in α-β stationary frame are expressed by:

$$\begin{cases} e_\alpha = \sqrt{\frac{3}{2}} E_m \cos(\omega t) \\ e_\beta = \sqrt{\frac{3}{2}} E_m \sin(\omega t) \end{cases} \tag{3.5}$$

The input voltage in the synchronous d-q coordinates (Fig. 3.6) are expressed by:

$$\begin{bmatrix} e_d \\ e_q \end{bmatrix} = \begin{bmatrix} \sqrt{e_\alpha^2 + e_\beta^2} \\ 0 \end{bmatrix} = \begin{bmatrix} \sqrt{\frac{3}{2}} E_m \\ 0 \end{bmatrix} \tag{3.6}$$

Input voltage in PWM rectifier The line-to-line input voltages of PWM rectifier can be described as:

$$\begin{cases} u_{ab} = (S_1 - S_2) V_{dc} \\ u_{bc} = (S_2 - S_3) V_{dc} \\ u_{ca} = (S_3 - S_1) V_{dc} \end{cases} \tag{3.7}$$

with S_{123} is the switching state of the converter, and V_{dc} is the DC-link output voltage.

The phase voltages are equal to:

$$\begin{bmatrix} v_{an} \\ v_{bn} \\ v_{cn} \end{bmatrix} = \frac{V_{dc}}{3} \begin{bmatrix} 2 & -1 & -1 \\ -1 & 2 & -1 \\ -1 & -1 & 2 \end{bmatrix} \begin{bmatrix} S_1 \\ S_2 \\ S_3 \end{bmatrix} \tag{3.8}$$

Model of three-phase PWM rectifier The voltage equations for balanced three-phase system without the neutral connection can be written as:

$$\begin{bmatrix} e_a \\ e_b \\ e_c \end{bmatrix} = \begin{bmatrix} v_{an} \\ v_{bn} \\ v_{cn} \end{bmatrix} + R \begin{bmatrix} i_a \\ i_b \\ i_c \end{bmatrix} + L\frac{d}{dt} \begin{bmatrix} i_a \\ i_b \\ i_c \end{bmatrix} \tag{3.9}$$

$$\mathbf{E}_{abc} = \mathbf{V}_{abc,n} + R\mathbf{I}_{abc} + L\frac{d}{dt}\mathbf{I}_{abc} \tag{3.10}$$

The current equation of the PWM controlled rectifier is defined as:

$$C\frac{dV_{dc}}{dt} + I_{dc} = S_1 i_a + S_2 i_b + S_3 i_c \tag{3.11}$$

Model of PWM rectifier in stationary coordinates $(\alpha - \beta)$ The voltage equation in the stationary α-β coordinates are written as:

$$\begin{bmatrix} e_\alpha \\ e_\beta \end{bmatrix} = \begin{bmatrix} v_\alpha \\ v_\beta \end{bmatrix} + R \begin{bmatrix} i_\alpha \\ i_\beta \end{bmatrix} + L\frac{d}{dt} \begin{bmatrix} i_\alpha \\ i_\beta \end{bmatrix} \tag{3.12}$$

and:

$$C\frac{dV_{dc}}{dt} + I_{dc} = S_\alpha i_\alpha + S_\beta i_\beta \tag{3.13}$$

with:

$$\begin{bmatrix} S_\alpha \\ S_\beta \end{bmatrix} = \frac{1}{\sqrt{6}} \begin{bmatrix} 2 & -1 & -1 \\ 0 & \sqrt{3} & -\sqrt{3} \end{bmatrix} \begin{bmatrix} S_1 \\ S_2 \\ S_3 \end{bmatrix} \tag{3.14}$$

Model of PWM rectifier in synchronous rotating coordinates $(d - q)$ The equations in the synchronous d-q coordinates are:

$$\begin{cases} e_d - v_d + R i_d + L\frac{di_d}{dt} - L\omega i_q \\ e_q = v_q + R i_q + L\frac{di_q}{dt} + L\omega i_d \end{cases} \tag{3.15}$$

and:

$$C\frac{dV_{dc}}{dt} + I_{dc} = S_d i_d + S_q i_q \tag{3.16}$$

where:

$$\begin{bmatrix} S_d \\ S_q \end{bmatrix} = \begin{bmatrix} \cos(\omega t) & \sin(\omega t) \\ \cos(\omega t) & -\sin(\omega t) \end{bmatrix} \begin{bmatrix} S_\alpha \\ S_\beta \end{bmatrix} \tag{3.17}$$

R can be practically neglected because voltage drop on resistance is much lower than voltage drop on inductance, what gives simplified equations of (3.9) (3.12), and (3.15).

The apparent power vector $\mathbf{S}_{\alpha\beta}$ can be expressed in several different manners as:

$$\mathbf{S}_{\alpha\beta} = \mathbf{E}_{\alpha\beta}.\mathbf{I}_{\alpha\beta}^*$$
$$= (e_\alpha + je_\beta)(i_\alpha - ji_\beta)$$
$$= (e_\alpha i_\alpha + e_\beta i_\beta) + j(e_\beta i_\alpha - e_\alpha i_\beta) \tag{3.18}$$
$$= P + jQ$$

where:

- $\mathbf{E}_{\alpha\beta}$ instantaneous power-source voltage vector,
- $\mathbf{I}_{\alpha\beta}$ instantaneous line current vector,
- $\mathbf{I}_{\alpha\beta}^*$ complex conjugate of vector $\mathbf{I}_{\alpha\beta}$,
- j imaginary unit.

The instantaneous active and reactive power supplied from the source is given by:

$$\begin{cases} P = e_\alpha i_\alpha + e_\beta i_\beta = e_a i_a + e_b i_b + e_c i_c \\ Q = e_\beta i_\alpha - e_\alpha i_\beta = \dfrac{1}{\sqrt{3}}(e_{bc} i_a + e_{ca} i_b + e_{ab} i_c) \end{cases} \tag{3.19}$$

It gives in the synchronous d-q coordinates:

$$\begin{cases} P = e_d i_d + e_q i_q \\ Q = e_q i_d - e_d i_q \end{cases} \tag{3.20}$$

For a unity power factor (Fig. 3.7), the following properties are considered:

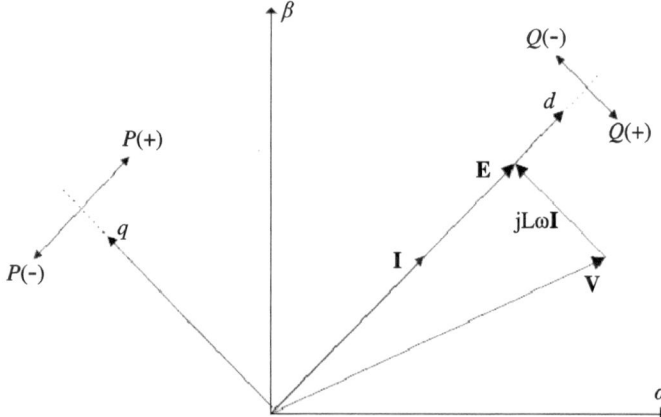

Fig. 3.7 Power flow in bi-directional AC/DC converter as dependency of **I** direction

$$\begin{cases} e_d = \sqrt{\frac{3}{2}} E_m \\ e_q = 0 \end{cases} \tag{3.21}$$

$$\begin{cases} i_d = \sqrt{\frac{3}{2}} I_m \\ i_q = 0 \end{cases} \tag{3.22}$$

3.4 Line Voltage and Virtual Flux Estimation

3.4.1 Line Voltage Estimator

An important requirement for a voltage estimator is to estimate the voltage correct also under unbalanced conditions and pre-existing harmonic voltage distortion. Not only the fundamental component should be estimated correct, but also the harmonic components and the voltage unbalance.

It is naturally possible to estimate the power-source voltage vector $\mathbf{E_{abc}}$ by simply adding converter output voltages $\mathbf{V_{abc,n}}$ to the voltage drops $\mathbf{V_{L_{abc}}}$ across the inductance L, which is calculated by the current differentiating, such that:

$$\mathbf{E_{abc}} = \mathbf{V_{abc,n}} + \mathbf{V_{L_{abc}}} \tag{3.23}$$

A second method utilizes powers P and Q as intermediate variables in estimating the power-source voltage vector $\mathbf{E_{abc}}$. Also, the estimated P and Q, given in Eq. (3.19), are used effectively as power feedback signals in the direct power controllers. However, Eq. (3.19) requires the power-source voltage $\mathbf{E_{abc}}$, which has to be eliminated to achieve the voltage sensorless operation.

Assuming that the line resistance R is negligible, and rewriting P and Q as a function of the switching state (S_1 S_2 S_3), the three-phase line currents (i_a, i_b, and i_c), the DC-bus voltage V_{dc}, and the inductance L, such that:

$$P = L \left(\frac{di_a}{dt} i_a + \frac{di_b}{dt} i_b + \frac{di_c}{dt} i_c \right) + V_{dc}(S_1 i_a + S_2 i_b + S_3 i_c) \tag{3.24}$$

$$Q = \frac{1}{\sqrt{3}} \left[3L \left(\frac{di_a}{dt} i_c - \frac{di_c}{dt} i_a \right) - V_{dc}[S_1 (i_b - i_c) + S_2 (i_c - i_a) + S_3 (i_a - i_b)] \right] \tag{3.25}$$

As can be seen in Eqs. (3.24) and (3.25), the estimating equations have to be changed according to the switching state (S_1 S_2 S_3) of the converter, and both equations require the parameter of the inductance L. However, this approach has the disadvantage that the current $\mathbf{I_{abc}}$ is differentiated, and noise in the current signal is gained through the differentiation.

Accounting for Eq. (3.18), the estimated line voltage vector $\mathbf{E_{\alpha\beta}}$ is expressed as:

$$\mathbf{E}_{\alpha\beta} = \frac{\mathbf{S}_{\alpha\beta}.\mathbf{I}_{\alpha\beta}}{||\mathbf{I}_{\alpha\beta}||^2} \tag{3.26}$$

which leads to:

$$\mathbf{E}_{\alpha\beta} = \begin{bmatrix} E_\alpha \\ \\ E_\beta \end{bmatrix} = \frac{1}{i_\alpha^2 + i_\beta^2} \begin{bmatrix} i_\alpha & -i_\beta \\ \\ i_\beta & i_\alpha \end{bmatrix} \begin{bmatrix} P \\ \\ Q \end{bmatrix} \tag{3.27}$$

$$\theta_{\mathbf{E}_{\alpha\beta}} = = \mathrm{tg}^{-1}\left(\frac{E_\beta}{E_\alpha}\right) \tag{3.28}$$

3.4.2   Virtual Flux Estimator

The voltage $\mathbf{E}_{\mathbf{abc}}$ imposed by the line power in combination with the three inductors RL are assumed to be quantities related to a virtual AC motor as shown in Fig. 3.8. Thus, R and L represent the stator resistance and the stator inductance of the virtual motor. The integration of the voltages $\mathbf{E}_{\mathbf{abc}}$, $\mathbf{V}_{\mathbf{abc,n}}$, and $\mathbf{V}_{\mathbf{L}_{\mathbf{abc}}}$ leads, respectively, to virtual flux vectors $\mathbf{\Phi}_{\mathbf{E}}$, $\mathbf{\Phi}_{\mathbf{V}}$, and $\mathbf{\Phi}_{\mathbf{V}_{\mathbf{L}}}$ in stationary α-β coordinates, as shown in Figs. 3.9 and 3.10.

Similarly to Eq. (3.23) a virtual flux equation can be presented as:

$$\mathbf{\Phi}_{\mathbf{E}} = \mathbf{\Phi}_{\mathbf{V}} + \mathbf{\Phi}_{\mathbf{V}_{\mathbf{L}}} \tag{3.29}$$

Based on the measured DC-link voltage V_{dc} and the converter switch states S_1, S_2, and S_3 the rectifier input voltages $\mathbf{V}_{\alpha\beta}$ are estimated as follows:

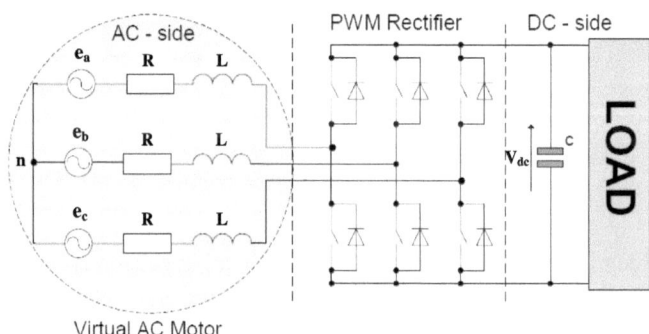

Fig. 3.8 Three-phase PWM rectifier with AC-side presented as virtual AC motor

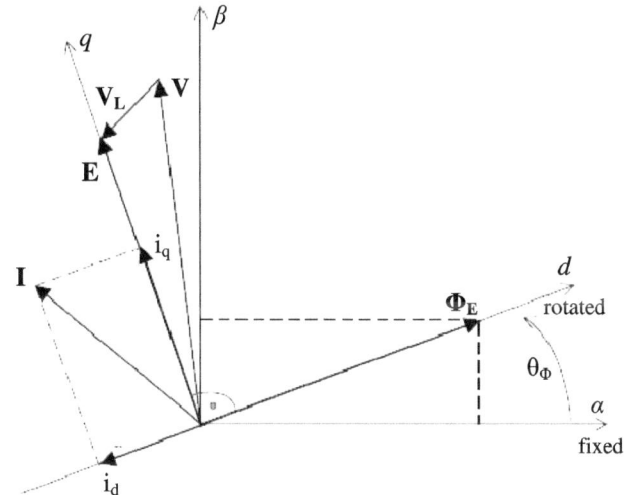

Fig. 3.9 Virtual flux vector $\mathbf{\Phi_E}$ aligned with the d-axis

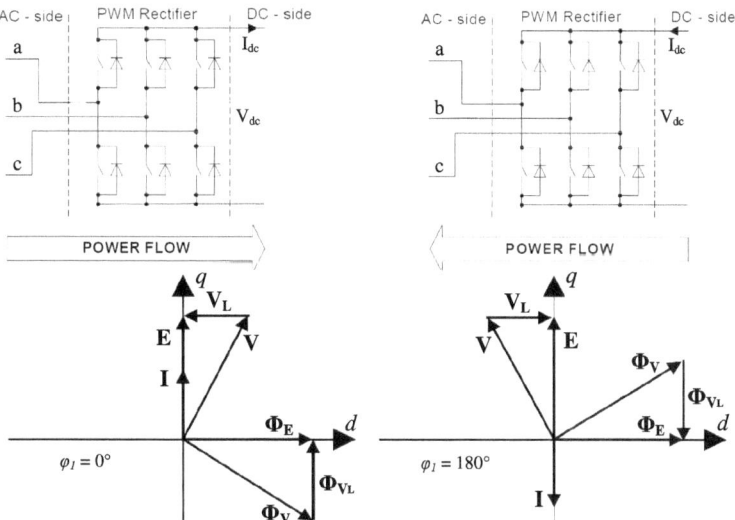

Fig. 3.10 Relation between voltage and flux vectors for different power flow direction in PWM rectifier

$$
\begin{cases}
v_\alpha = \sqrt{\frac{2}{3}} \left(S_1 - \frac{1}{2}(S_2 + S_3) \right) V_{dc} \\
v_\beta = \frac{1}{\sqrt{2}} (S_2 - S_3) V_{dc}
\end{cases}
\tag{3.30}
$$

Then, the line virtual flux components of vector $\mathbf{\Phi_E}$ are calculated from Eq. (3.30) in stationary α-β coordinates system as follows:

$$
\begin{cases}
\phi_{E_\alpha} = \int (v_\alpha + L\frac{di_\alpha}{dt})dt = \int v_\alpha dt + Li_\alpha \\
\phi_{E_\beta} = \int (v_\beta + L\frac{di_\beta}{dt})dt = \int v_\beta dt + Li_\beta
\end{cases}
\tag{3.31}
$$

If the resistance R is taken into account, Eq. (3.31) becomes:

$$
\begin{cases}
\phi_{E_\alpha} = \int (v_\alpha + Ri_\alpha)dt + Li_\alpha \\
\phi_{E_\beta} = \int (v_\beta + Ri_\alpha)dt + Li_\beta
\end{cases}
\tag{3.32}
$$

The position of the virtual flux vector can be determined accounting for the angle $\theta_{\mathbf{\Phi_E}}$:

$$
\theta_{\mathbf{\Phi_E}} = tg^{-1} \left(\frac{\phi_{E_\beta}}{\phi_{E_\alpha}} \right)
\tag{3.33}
$$

By considering the following relation:

$$
\begin{cases}
\phi_{E_\alpha} = \int e_\alpha dt = \frac{e_\beta}{\omega} \\
\phi_{E_\beta} = \int e_\beta dt = -\frac{e_\alpha}{\omega}
\end{cases}
\tag{3.34}
$$

Equation (3.19) leads to:

$$
\begin{cases}
P = \omega(\phi_{E_\alpha} i_\beta - \phi_{E_\beta} i_\alpha) \\
Q = \omega(\phi_{E_\alpha} i_\alpha + \phi_{E_\beta} i_\beta)
\end{cases}
\tag{3.35}
$$

3.5   Implementation of DPC Strategy

3.5.1   Basic Block Diagram of DPC

The main idea of DPC strategy is similar to the well-known Direct Torque Control (DTC) for AC motors. Instead of torque and stator flux, the instantaneous active P and reactive Q powers are controlled. Figure 3.11 shows the basic block diagram of DPC.

The references of active power P^* (delivered from the outer DC voltage controller) and reactive power Q^* (set to zero for unity power factor) are compared with the estimated P and Q values. The errors between the references and the estimated values are the input signals

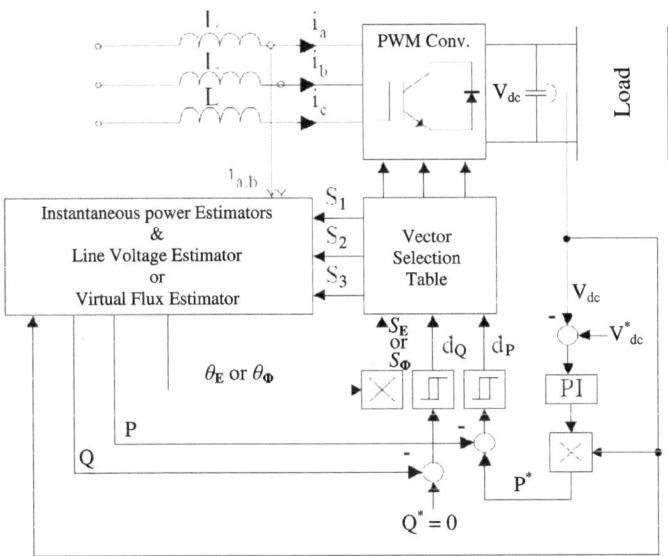

Fig. 3.11 Implementation scheme of DPC strategy for three-phase PWM rectifier

of the two-level hysteresis controllers. These controllers generate two control signals d_P and d_Q, such that:

- $d_P = +1$ for $P < P^* - \varepsilon_P$,
- $d_P = -1$ for $P > P^* + \varepsilon_P$.

and:

- $d_Q = +1$ for $Q < Q^* - \varepsilon_Q$,
- $d_Q = -1$ for $Q > Q^* + \varepsilon_Q$.

where $\pm \varepsilon_P$ and $\pm \varepsilon_Q$ are the hysteresis bands.

The control signals (d_P and d_P) and the voltage vector position S_E or flux vector position S_Φ are the input signals of the look-up table (vector selection table), which selects the appropriate combination of the IGBT states.

As illustrated in Fig. 3.12a, the α-β plane is divided into twelve sectors, and the sector S_E defining the position of the line voltage **E** can be numerically expressed as:

$$\text{Sector } n: \ (n-2)\frac{\pi}{6} \ \leq \ \theta_E \ < \ (n-1)\frac{\pi}{6}; \quad n = 1, 2, \ldots, 12 \tag{3.36}$$

As shown in Fig. 3.12b, the sector S_Φ defining the position of the virtual flux Φ_E can be numerically expressed as:

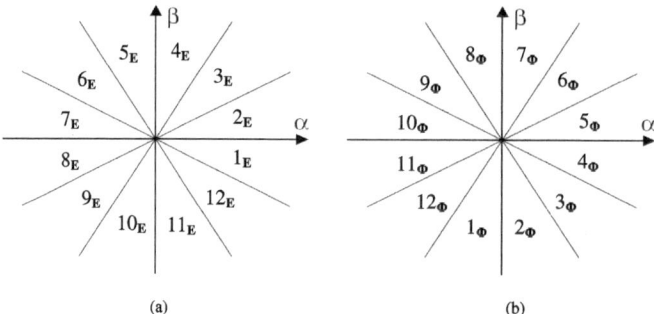

(a)   (b)

Fig. 3.12 Sector selection for **a** V-DPC strategy and **b** VF-DPC strategy

$$\text{Sector } p: \ (p-5)\frac{\pi}{6} \ \leq \ \theta_E \ < \ (p-4)\frac{\pi}{6}; \quad p = 1, 2, \dots, 12 \qquad (3.37)$$

3.5.2   Vector Selection Table of DPC

The instantaneous active power P and reactive power Q depend on the positions of the line voltage vector **E** and of the applied converter voltage vector **V**.

The applied converter voltage vector **V** has indirect influence on inductance voltage vector $\mathbf{V_L}$ as well as phase and amplitude of line current vector **I**.

According to Eq. (3.20), if the line voltage vector **E** is aligned with the d-axis, the active and reactive powers are expressed as follows:

$$\begin{cases} P = e_d\, i_d \\ Q = -e_d\, i_q \end{cases} \qquad (3.38)$$

The variations of the powers during each sampling period T_s depend on the variation of the d- and q- components of the line current vector **I**, such that:

$$\begin{cases} \Delta P = e_d\, \Delta i_d \\ \Delta Q = -e_d\, \Delta i_q \end{cases} \qquad (3.39)$$

From Eq. (3.39), one can noted that:

$$\begin{cases} P \Uparrow \ \Longrightarrow \ \Delta P > 0 \ \Longrightarrow \ \Delta i_d > 0 \ \Longrightarrow \ i_d \Uparrow \\ P \Downarrow \ \Longrightarrow \ \Delta P < 0 \ \Longrightarrow \ \Delta i_d < 0 \ \Longrightarrow \ i_d \Downarrow \end{cases} \qquad (3.40)$$

and:

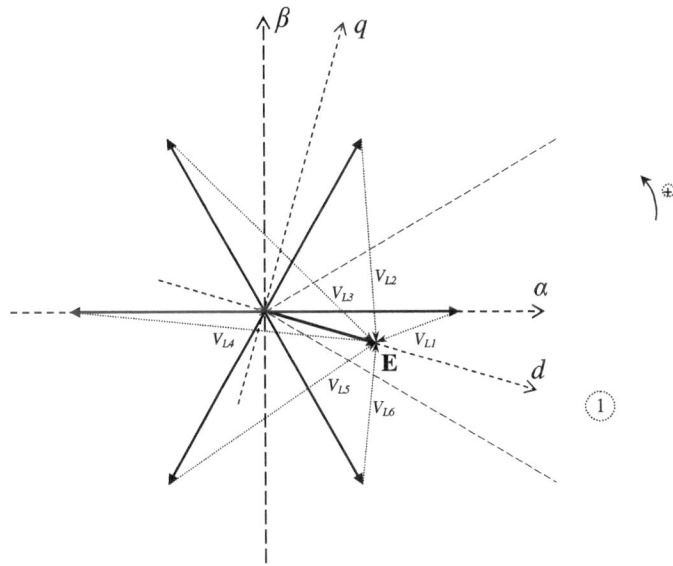

Fig. 3.13 Instantaneous position of vectors for the case where the line voltage vector **E** is located in sector 1

Table 3.1 Effect of the converter voltage vectors **V** on active power P and reactive power Q for the case where the line voltage vector **E** is located in sector 1

Sector 1$_E$	$\mathbf{V_1}$	$\mathbf{V_2}$	$\mathbf{V_3}$	$\mathbf{V_4}$	$\mathbf{V_5}$	$\mathbf{V_6}$	$\mathbf{V_{0,7}}$
P	⇓	⇑⇑	⇑⇑	⇑⇑	⇑⇑	⇓	⇑⇑
Q	⇑	⇑⇑	⇑⇑	⇓	⇓⇓	⇓⇓	⇌

$$\begin{cases} Q \Uparrow \implies \Delta Q > 0 \implies \Delta i_q < 0 \implies i_q \Downarrow \\ Q \Downarrow \implies \Delta Q < 0 \implies \Delta i_q > 0 \implies i_q \Uparrow \end{cases} \quad (3.41)$$

For each sampling period T_s, one can noted that:

$$\mathbf{V_L} = L\frac{d\mathbf{I}}{dt} = L\frac{\Delta\mathbf{I}}{T_s} = \frac{L}{T_s}\Delta i_d + j\frac{L}{T_s}\Delta i_q = V_{L_d} + j\, V_{L_q} \quad (3.42)$$

Therefore, the variations of the d- and q- components of the inductance voltage vector $\mathbf{V_L}$ can describe the evolutions of the active and reactive powers.

Figure 3.13 shows the inductance voltage vectors $\mathbf{V_L}$ generated with the application of the converter voltage vectors **V**, for the case where the line voltage vector **E** is located in sector 1. Referring to Fig. 3.13, the variations of the d- and q- components of the inductance voltage vector $\mathbf{V_L}$ gives the evolutions of the active power P and reactive power Q as depicted in Table 3.1.

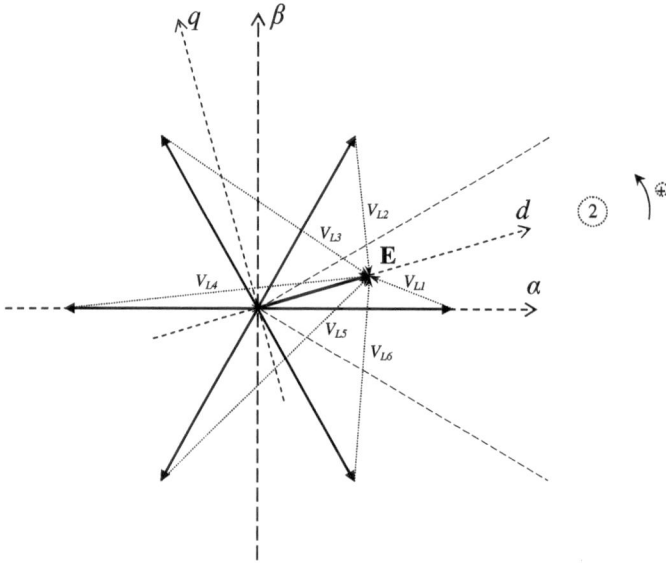

Fig. 3.14 Instantaneous position of vectors for the case where the line voltage vector **E** is located in sector 2

Table 3.2 Effect of the converter voltage vectors **V** on active power P and reactive power Q for the case where the line voltage vector **E** is located in sector 2

Sector $2_\mathbf{E}$	$\mathbf{V_1}$	$\mathbf{V_2}$	$\mathbf{V_3}$	$\mathbf{V_4}$	$\mathbf{V_5}$	$\mathbf{V_6}$	$\mathbf{V_{0,7}}$
P	⇓	⇓	⇑⇑	⇑⇑	⇑⇑	⇑⇑	⇑⇑
Q	⇓	⇑⇑	⇑⇑	⇑	⇓⇓	⇓⇓	⇌

Figure 3.14 shows the different inductance voltage vectors $\mathbf{V_L}$ generated with the application of the converter voltage vectors **V**, for the case where the line voltage vector **E** is located in sector 2. Referring to Fig. 3.14, the variations of the d- and q- components of the inductance voltage vector $\mathbf{V_L}$ gives the evolutions of the active power P and reactive power Q as depicted in Table 3.2.

On the basis of Tables 3.1 and 3.2, several switching table have been proposed in literature. Tables 3.3 and 3.4 describe two vector selection tables of DPC strategy for PWM converter.

3.6   Conclusion

In conclusion, Direct Power Control proves to be a powerful approach for controlling induction motor drives, offering significant advantages in terms of dynamic response and reduced complexity compared to traditional control methods. The direct handling of active and

Table 3.3 First vector selection table of DPC strategy for PWM converter

d_P	+1	+1	−1	−1
d_Q	+1	−1	+1	−1
Sector 1	3	5	1	6
Sector 2	3	5	2	1
Sector 3	4	6	2	1
Sector 4	4	6	3	2
Sector 5	5	1	3	2
Sector 6	5	1	4	3
Sector 7	6	2	4	3
Sector 8	6	2	5	4
Sector 9	1	3	5	4
Sector 10	1	3	6	5
Sector 11	2	4	6	5
Sector 12	2	4	1	6

Table 3.4 Second vector selection table of DPC strategy for PWM converter

d_P	+1	+1	−1	−1
d_Q	+1	−1	+1	−1
Sector 1	2	4	1	6
Sector 2	3	5	2	1
Sector 3	3	5	2	1
Sector 4	4	6	3	2
Sector 5	4	6	3	2
Sector 6	5	1	4	3
Sector 7	5	1	4	3
Sector 8	6	2	5	4
Sector 9	6	2	5	4
Sector 10	1	3	6	5
Sector 11	1	3	6	5
Sector 12	2	4	1	6

reactive power enables DPC to provide rapid adjustments with minimized power and torque ripple, enhancing both efficiency and system stability. However, DPC's performance can be affected by parameter variations, particularly in power-source voltage estimation, which may introduce control challenges under dynamic load conditions. Future research could focus on adaptive estimation techniques and integrating DPC with predictive control strategies to further improve robustness and efficiency for a wide range of industrial applications.

References

1. Alonso-Martínez, J., Carrasco, J. E. G., & Arnaltes, S. (2010). Table-based direct power control: A critical review for microgrid applications. *IEEE Transactions on Power Electronics, 25*(12), 2949–2961.
2. Vazquez, S., Sanchez, J. A., Carrasco, J. M., Leon, J. I., & Galvan, E. (2008). A model-based direct power control for three-phase power converters. *IEEE Transactions on Industrial Electronics, 55*(4), 1647–1657.
3. Vazquez, S., Sanchez, J. A., Carrasco, J. M., Leon, J. I., & Galvan, E. (2008). A model-based direct power control for three-phase power converters. *IEEE Transactions on Industrial Electronics, 55*(4), 1647–1657.
4. Gui, Y., Lee, G. H., Kim, C., & Chung, C. C. (2017). Direct power control of grid connected voltage source inverters using port-controlled Hamiltonian system. *International Journal of Control, Automation and Systems, 15*, 2053–2062.
5. Yessef, M., Bossoufi, B., Taoussi, M., Benbouhenni, H., Lagrioui, A., & Chojaa, H. (2023). Intelligent direct power control based on the neural super-twisting sliding mode controller of a DFIG. In *International conference on digital technologies and applications* (pp. 726–735). Cham: Springer Nature Switzerland.
6. Krama, A., Zellouma, L., Benaissa, A., Rabhi, B., Bouzidi, M., & Benkhoris, M. F. (2019). Design and experimental investigation of predictive direct power control of three-phase shunt active filter with space vector modulation using anti-windup PI controller optimized by PSO. *Arabian Journal for Science and Engineering, 44*(8), 6741–6755.
7. Kadem, M., Semmah, A., Wira, P., & Dahmani, S. (2020). Fuzzy logic-based instantaneous power ripple minimization for direct power control applied in a shunt active power filter. *Electrical Engineering, 102*(3), 1327–1338.

.